From Arcade Street to Main Street

From Arcade Street *to* Main Street

A HISTORY OF THE SEEGER REFRIGERATOR COMPANY

1902–1984

James B. Bell

FOREWORD BY
George Latimer
Former Mayor of St. Paul

Ramsey County Historical Society
ST. PAUL, MINNESOTA

Production Credits

Copyediting: Virginia Brainard Kunz
John M. Lindley, John M. Lindley & Associates
Text Design: Wendy Holdman, Stanton Publication Services
Composition: Stanton Publication Services
Cartography: Patti Isaacs, Parrot Graphics
Dust Jacket Design: Kyle Hunter, Stanton Publication Services
Indexing: John M. Lindley, John M. Lindley & Associates
Production Management: John M. Lindley, John M. Lindley & Associates
Printing and Binding: Sexton Printing

Published by
Ramsey County Historical Society
323 Landmark Center
75 West Fifth Street
St. Paul, MN 55102
(651) 222-0701
www.rchs.com

Printed in the United States of America.

ISBN: 978-0-934294-67-6

To the Memory of
John Augustus Seeger,
Walter G. Seeger, and Gustave R. Seeger
and the Thousands of Men and Women
Whose Talents, Skills, and Loyalty to the
Seeger Refrigerator Company
Produced Extraordinary Achievements

Contents

Foreword

CHRONICLED HERE IS THE RISE of the Seeger Corporation during the first part of the century leading to its merger with Whirlpool in 1955, and its demise with the shutting down of the St. Paul plant by the Whirlpool Corporation in 1984. During the decade of the 1980s, St. Paul suffered a hollowing out of its manufacturing and transportation base. Joining Whirlpool, American Hoist and Derrick, Brown and Bigelow, Burlington Northern, and West Publishing Company—all shut down or moved out of the city after nearly a century of providing jobs for the workingmen and women of Minnesota's capital city. And many of the long-time employers who did not move out, reduced their skilled and semiskilled work force through automation.

Jerry Ficocello of St. Paul, for example, joined the *Pioneer Press* newspaper as one of 400 typesetters in 1963. When he retired in 2000, only 30 typesetters remained. Obviously, this was a huge increase in productivity, but also a decline in job opportunities for skilled and semiskilled workers. Professor Bell documents the same increase in productivity of the St. Paul Whirlpool plant. From 1956, the date of the merger, to 1968, the numbers of units shipped rose from 180,795 to 916,000 while the number of employees did not rise.

Professor Bell, through diligent research impeded by the sad destruction of records at the time of the plant's closing, documents how the Seeger Company flourished in the earlier part of the 20th century through the leadership of the Seeger family. At one level the Seeger story mirrors the deindustrialization of America. This has been documented by others, including William Julius Wilson in *When Work Disappears: The World of the New Urban Poor* (New York, 1996). What distinguishes Professor Bell's coverage is his intimate portrait of the social and economic benefits during the growth of Seeger and the calamitous personal and social effects of the Whirlpool closing. Especially poignant, were his numerous interviews with long-time employees who were facing a bleak job landscape after devoting most of their work life to Whirlpool-Seeger. The author notes that of the 625 production line workers still with the company at the time of closing, the average age was 47 with an average tenure of 22 years. They stayed because they did well by working hard in a culture that set high expectations and rewarded accordingly. Many of these workers I knew personally because as a young lawyer prior to my election as mayor of St. Paul, I represented their union, Local 827. One employee was quoted as saying, "Like Seeger before it, Whirlpool was considered a source of secure employment and decent wages on the East Side. It has been around for 80 years. If you didn't finish high school or didn't go off to college, you worked at Whirlpool. Some families are on their third generation there." Bob Schwan was a union officer whom I knew and the author accurately quoted Schwan's pride in the union acting as a good middleman between the workers and the company.

This collegiality was set by John Seeger who walked through the plant every morning and greeted the employees, many of them by first name. The loyalty to their company survived even the bitterness of the closing. I believe it is because the culture of expectations, respect, and reward were long set by the Seeger leadership. It is no wonder that as a young lawyer I would hear workers tell me that they would meet me at the Seeger plant some twenty years after the plant was Whirlpool's.

This collegiality reflected community values as well as the Seeger family's. In her excellent history of St. Paul, *Claiming the City: Politics, Faith, and the Power of Place in St. Paul* (Ithaca, N.Y., 2001), Mary Lethert Wingerd notes that the language of civic responsibility was shared across the bargaining table. She features a photograph of H.B. Fuller, who like Seeger was the founder of a family-owned business that flourished as a public corporation in the twentieth century. H.B. Fuller is shown seated on a log, surrounded on either side by his employees. It is not a stretch to imagine the Seegers in a similar photograph with their employees.

But the earth had gone flat for workers long before Thomas Friedman discovered it. I have no independent recollection of the business plans which motivated Whirlpool's decision to shut down the Saint Paul plant. I do know that the East Side of Saint Paul is still struggling to recover from that crushing event. Had the Seegers maintained control, the outcome might well have been different.

George Latimer
February 2007

Preface

FOR MORE THAN 100 YEARS in St. Paul, Minnesota, the name Seeger was synonymous with cabinetmaking and icebox, refrigerator, and freezer manufacturing. The origins of the Seeger Refrigerator Company in 1902 were modest and its future uncertain. Yet the purposeful vision of its founder, John Augustus Seeger, guided the fledgling enterprise with a steady hand from its establishment until his death in 1938. His grandson and a later president of the firm, John S. Holl, remembered him as "the King," the man in charge.

Setting an example for all of his employees, John A. Seeger was at his office every morning before eight o'clock. His first task of the day was to walk through the plant and say his "good mornings" to the workers, many of whom he knew by their first names. The tour also gave him an opportunity to note any problems that might be emerging on the assembly line or in any of the other departments. The company became a vital and worthy economic force in both the St. Paul region and nationally while the family members who directed the destiny of the firm played significant roles in local civic affairs. From its modest beginnings with three employees in 1902, the firm grew to a work force of more than 6,100 persons during the 1950s. The leadership of such family members as the founder, John A. Seeger (1853–1938), and his two sons Walter G. Seeger (1886–1969) and Gustave R. Seeger (1889–1945), along with the contributions of a cavalcade of thousands of exceptionally talented, skilled, loyal, and long-serving employees advanced the company for much of the twentieth century. The Seeger story is the stuff of which the American experience has been fashioned. It is the story of the son of German immigrants and his persistent pursuit of economic and civic accomplishment.

From Arcade Street to Main Street examines and describes the history of a family-held and -led company from its founding in 1902 and traces its development through the following decades to its merger with two other corporations in 1955 to form the Whirlpool-Seeger Corporation. An overview of the successor corporation's experience in St. Paul until 1984 is also presented when the last vestige of the Seeger manufacturing legacy was closed.

The story of the Seeger Company is intertwined with several forces. First, it is at once a chronicle of an early Minnesota family and of three generations of its members who played a strong role in the firm's advancement and success. It is an account that is steeped too in the fabric of St. Paul's and Minneapolis's social and economic history for nearly the entire twentieth century. While in the late nineteenth century nearly every United States community of any size had cabinetmakers that built iceboxes, the Seeger Company rode the crest of the movement locally, regionally, and nationally to provide refrigerated appliances for American homes and businesses. The technological and design advancements of the refrigeration industry and the advent of the expansion of electrical services in the 1920s allowed Seeger to manufacture products that contributed to the modernization of homes and businesses with mechanical refrigerators. The experience is a chronicle of the twentieth century and reflects in part the transformation of the refrigeration industry: a story that has not been previously recounted for Seeger, Whirlpool-Seeger, or the modern-day Whirlpool corporations. The company's story is engaging, diverse, and complicated. My design as a historian has been to provide an interpretation that is inclusive and comprehensive rather than exclusive and parochial in substance and scope. To provide an account of the course of the Seeger Company's experience, I have raised again the well-worn and simple questions such as: "What happened?" "How did it happen?" and "When did it happen?" My aimed has been to illuminate and understand a corporation's unfolding accomplishments and purpose through the years.

The roles that Seeger, the Whirlpool Corporation, and the RCA Corporation played in providing an essential foundation for the merged Whirlpool-Seeger Corporation in 1955 is an untold story. It is an account, based on historical records and personal interviews with many former employees of the Seeger, Whirlpool-Seeger, and the Whirlpool corporations. The historical account is at substantial variance with the occasional descriptions of the company's story broadcast by one corporate official or another or printed in various publications.

In the course of my research efforts I have had the opportunity and enriching personal pleasure to interview numerous former employees of Seeger and its successor companies, Whirlpool-Seeger, and the Whirlpool Corporation. Through my discussions with these individuals I have learned about the daily routine of the Arcade Street factory. Their recollections focused on a variety of experiences: such as the hours worked each day, the beginning wage, the breaks for coffee or lunch, or the relentless requirements and rhythms of the assembly line, and the friendships they made with colleagues who also worked at the St. Paul plant. Memories punctuated and embellished by the strengths and weaknesses of fellow employees, by company-sponsored sports activities, picnics, Christmas parties, and educational opportunities for their sons and daughters.

A manufacturing plant, such as the Seeger Refrigerator Company, is at bottom a social organization. It employs a certain number of persons who usually work shifts of about eight hours a day, five days a week throughout the year, spending about one-third of their lives on the job. At Seeger it was a common experience for employees to work in various departments 20, 25, 30 or more years contributing to the manufacturing of refrigerators, freezers, milk coolers, and other products. Today, such an association would be classified as an extraordinary demonstration of corporate loyalty.

The workplace gave rise to a process of bonding and friendship that created an exceptional sense of loyalty among the company's employees. Through the years they have supported one another as they have encountered the inevitable irregularities of life. They were together on those special occasions that mark the life experience, such as at baptisms or weddings of sons and daughters, retirement parties, and funerals. Out of the many interviews I conducted, I found that in a vital and rich sense a good number of Seeger employees felt their life was their job and their job was their life. Thus the lives of these employees were intertwined with the evolving successful fortune of the company.

Yet, I must report that there are gaps in the historical account that will never be filled. The situation is compounded by the lack of a corporate archive containing the company's records for the half-a-century before the 1955 merger of Seeger with the Whirlpool Corporation and the Appliance Division of the Radio Corporation of America to form the Whirlpool-Seeger Corporation. A situation created by the massive and catastrophic destruction of the Seeger Refrigerator Company's historical records when the plant was closed in December 1984.

Former employees have recounted to me that they witnessed the disposal of enormous quantities of manuscript and printed materials relating to Seeger's history between 1902 and 1955. The destruction and disposal of such materials as ledger and journal books, blueprints of products and parts manufactured at the plant, catalogs of products manufactured by the company for more than fifty years, correspondence files, photographs of the factories and operations in St. Paul and Evansville, Indiana, and so forth. Lost with one destructive stroke were the records of a company that played an important role in the history of the refrigeration industry in the United States, and in the business life of St. Paul and Minnesota for nearly a century.

The absence of the records impedes the historian to provide a thorough and systematic examination of various aspects of the company's history. For example, when John A. Seeger founded the company to manufacture iceboxes in 1902 there were two additional employees. By the early 1920s several hundred individuals worked at the plant and by the beginning of the 1950s more than 6,100 persons were employed at the facilities in St. Paul, Minnesota, and at Evansville in Indiana. Without an archive containing Seeger personnel records, it is impossible to recover the names of a vast majority of the men and women who were associated with the company over more than half-a-century. We are unable to know where they lived, how many years they were employed, the kinds of jobs they performed, and the departments of the factory with which they were associated.

During the course of research for this book I learned that frequently several members of a family, sometimes over two or three generations, were employed at the Seeger factory. Unfortunately the destruction of the company's personnel records does not allow for the reconstruction of such significant associations. Furthermore, it is commonly asserted that many workers resided in the various neighborhoods on St. Paul's East Side and it would be a fascinating historical study to examine this subject and determine the areas of residence of the workers in St. Paul, Minneapolis, and the surrounding region. From the beginning of St. Paul's history, the community was comprised of people representing a background of diverse ethnic and national origins. Because Seeger employed thousands of workers between 1902 and 1955, it would be interesting to know their origins and whether they were native-born or from countries in Europe, Africa, Latin America, and elsewhere.

Another example of what has been lost through the destruction of the Seeger corporate records is the systematic and comprehensive evidence that traces the launch, development, and accomplishments of Seeger's Sales Department. From the company's beginnings in the first decade of the twentieth century a strategy was developed and implemented by John A., Walter G., and Gustave R. Seeger to encourage the sales of its products. When Charles Muessel joined the firm in 1909, he aggressively expanded the promotional program. Catalogs of the company's wide range of products were regularly published, advertisements appeared in key national magazines that enjoyed an informed and affluent readership, and sales offices were opened in major cities across the United States. While records of the Whirlpool Corporation's short-lived operations in St. Paul may have been transferred to the corporate headquarters in Benton Harbor, Michigan, the records of the Seeger Company were not.

The history of the Seeger Refrigerator Company falls into several identifiable stages. The first period was dominated by the manufacturing of iceboxes for domestic and commercial use and the first years of transition to electrical refrigerators and covers the years between 1902 and 1928. It was followed by the era that witnessed Seeger's initial alliance with Sears, Roebuck and Company to manufacture the *Coldspot* refrigerator, running from about 1930 until the outbreak of World War II. During this period the firm became a major national manufacturer of refrigerators applying complex technological equipment, assembly-line production procedures, and an expanded force of professional and skilled employees. The next phase of the company's history was Seeger's extensive role in manufacturing a variety of items for the use of the army, navy, and air force during World War II and the Korean War. The conversion to peacetime manufacturing in the last half of the 1940s saw a significant increase in the company's sales and productivity between 1945 and 1955. The decision to merge with the Whirlpool Corporation and the RCA Appliance Division in 1955 is then explained in detail. This account of the Seeger Company concludes with an examination of the successor corporation, which was known for two years as the Whirlpool-Seeger Corporation (1955–1957), and the final years of the post-1957 Whirlpool Corporation's presence in St. Paul (1957–1984), which was centered on the original Seeger factory site on Arcade Street.

James B. Bell

Acknowledgments

AMONG THE MANY PERSONS whom I have interviewed and who shared with me their extensive knowledge about the complex manufacturing operations at Seeger, I am particularly indebted to the following persons: Dwain Bell, Darrell Borgstrum, Paul Bowman, Eilef Bradford, Donald Casey, Clinton Conrad, John Diesslin, Gene Gardner, Harry Gauld, Arthur Gehring, Raul Gomez, Richard Hagberg, Edward Hartung, Orrie Heasley, John Herman, John O. Jahnke, Robert Johnson, Robert Kelly, Louis Kuchenmeitster, William Littler, Mary Turner Maher (Philip Turner), Gerald Nielson, Ray Phillips, Howard Pittelkow, Joyce Rensch, George Ricci, Harland and Rosemary St. George, Col. Dennis M. Savage (Ret.), Harvey Schultz, Brant Short, Harvey Stoltz, Frank Taylor, Floyd Wefring, Clarence Williams, Audrey H. Winkler (Wynn Winkler).

I am also grateful to Eileen McCormack, Curator of Manuscripts of the James J. Hill Reference Library, who brought to my attention a variety of materials relating to John A. Seeger's business relationships with James J. Hill and his son Louis W. Hill. I am indebted to Jane Hopeman, reference librarian of the William Mitchell College of Law, for her knowledgeable assistance in tracing the history of the Seeger Refrigerator Company's long-time litigation with the Bohn Refrigerator Company.

Many former employees of the Seeger and Whirlpool corporations read all or part of the manuscript for this book including Don Casey, Clinton Conrad, Richard Hagberg, Gene Gardner, Ray Phillips, Herb Rhoda, Brant Short, Floyd Wefring, and Clarence Williams. All of them contributed invaluable comments and suggestions that have saved me from many errors and provided informative details and insights regarding the company's operations.

A number of persons personally associated and interested with the long history of the Seeger Refrigerator Company generously supported the publication of this volume and to each of them I convey my heartfelt thanks. The book would not have reached the printer's bench or the reader's hands without the timely assistance of several anonymous Seeger family members, James and Ursula Andrews, Walter and Melinda Andrews,

Mrs. Alan Buckbee, Jack C. and Martha H. Foote, Mr. Thomas H. Garrett III, Mrs. Shirley Muessel Jordan, Terry and Ann Love, Mr. and Mrs. Charles A. Muessel, Jr., Janet Reay Walton and James T. Walton, Hugh and Marilyn Wooldridge, the Right Reverend Robert M. Wolterstorff, and Walter S. and Barbara Wolterstorff.

For more than half-a-century the Murphy Companies of St. Paul and the Seeger Refrigerator Company maintained a business association. I am grateful to Laurel May Murphy for her gracious encouragement and support toward the publication of this book. My colleague and historian of St. Paul's and Minnesota's past, Richard H. Nicholson, has cheerfully and informatively shared with me his house at 260 Summit Avenue in St. Paul in which John A. Seeger's firm supplied the millwork and cabinetry.

The Seeger Refrigerator Company was founded and flourished in St. Paul, Ramsey County, Minnesota. Led by John A. Seeger and in turn by his sons Walter and Gustave, each of the company officials played prominent roles in the community's corporate and civic life for more than six decades. As the author of this book, I felt it would be appropriate for this history of a long-time local employer to be published by a St. Paul organization committed to preserving and interpreting the region's story.

A joint effort to accomplish the task was forged with the Ramsey County Historical Society, an agency that is dedicated to helping a community remember itself—the events, issues, institutions, and persons that experienced and shaped its history. I have been fortunate on several counts to be associated with the Society in the publication of this chronicle of the Seeger Refrigerator Company. In particular the book, over the years of research and writing, has benefited from the talents and skills of two outstanding professional editors, Virginia Brainard Kunz and John M. Lindley. Both editors have taken a keen interest in the book's development and progress toward the printer's bench. Their extraordinary talents and expertise were individually distinctive but mutually complementary. Without reservation I readily admit that the book has been the recipient of generous and tactful editorial guidance and an impressive reservoir of knowledge relating to St. Paul's history.

But writing and publishing a book is not a one-person show. It requires the thoughtful efforts of many individuals. I am indebted to other members of the Society's staff including the Executive Director, Priscilla Farnham, who has been a steady source of encouragement at every stage of the book's development. Maureen McGinn and Jana Armstead brought prompt and helpful replies to endless requests for research assistance on a variety of matters. The distinguished members of the Society's Editorial Board have also offered sustained interest in the project. I am indeed grateful to each and all of these individuals for their encouragement and support.

An earlier version of Chapters Two and Three appeared in *Ramsey County History,* vol. 30, no. 1 (Spring 1995), a publication of the Ramsey County Historical Society, St. Paul, Minnesota.

Unless otherwise noted, the citations in the endnotes and the photo captions in this book refer to materials in the Collection of James B. Bell of the Seeger Refrigerator Company, the Whirlpool-Seeger Corporation, and the Whirlpool Corporation.

From Arcade Street to Main Street

1

1856: The Seeger Family Arrives in Minnesota

IT IS UNCLEAR WHEN the Seeger family arrived in America from what is today Germany, nor is it known why they immigrated. We do know that there was widespread discontent and revolution in Germany in 1848 and 1849, first in the south and west, then in the central states and Prussia. There were public demonstrations and demands for constitutional changes and reform. Perhaps the family fled their homeland in the wake of the political turmoil and the prospect that their sons would inevitably be obliged to face military service. In any event by 1852 the Seeger family was residing in Covington, Kentucky, where John A. Seeger was born in 1853. He arrived with his family in Minnesota, on October 1, 1856, settling first near St. Peter, in the Minnesota River valley. John's father, William Seeger, helped to locate a colony of German settlers at New Ulm in that year. He established a brewery soon after his arrival and later engaged in the mercantile business for five years in Le Seuer. These were years of unrest for the Dakota people along the Minnesota River frontier, and the Seeger family probably experienced first-hand, as did their neighbors in the area, the uncertainties of the 1862 conflict.

The family moved to St. Paul in 1868 and William Seeger became deputy treasurer of the state of Minnesota. His appointment to the office was perhaps partly based on his service recruiting Germans for settlement in Minnesota, or possibly as a consequence of his Civil War experience as a sergeant in Company D, of the Ninth Regiment of Minnesota's Volunteer Infantry, or most likely because his daughter Bertha had married the state treasurer, Emil Munch Jr. of Stillwater in 1865. Seeger served in the post until 1872 when he briefly succeeded his son-in-law as state treasurer and was caught in the deadly cross-fire of bitterly partisan politics in the legislature.[1]

Thirteen years old when he moved to St. Paul with his family, John A. Seeger went to work for Charles Leonard, a Massachusetts native and a cabinetmaker and carpenter who had arrived in the town in 1858. In 1868, at the age of thirteen, he began to serve his apprenticeship as architect, cabinetmaker, and carpenter with the Leonard-Shiere Company, a well-known contractor of the city. Completing his apprenticeship at the age of sixteen, he continued on with the company. Within a few years he became a member of the firm known as Leonard & Seeger Company, contractors and builders, with a shop at 145 Jackson Street. His ability and maturity were demonstrated in the building of the Assumption Church in 1871 when the young architect was called upon to translate the plans of the church, which had been made in Germany, to older workmen. One of Seeger's early projects was participating in the construction between 1871 and 1874 of the twin–towered, modified Romanesque–styled, Assumption Church on today's West Seventh Street. It was then, as it is today, a landmark on the skyline

The Major Players in the Seeger Story

Anthony Yoerg and His Brewery

Anthony Yoerg launched Minnesota's first brewery in 1848 in his house at Eagle and Washington streets in the Upper Landing neighborhood of St. Paul. Born in 1816 in Gundelfingen, on the Brenz River about twenty miles northeast of the city of Ulm in Bavaria, he came to the United States about 1845 and settled in Pittsburgh before removing to St. Paul two years later.

Anthony Yoerg, about 1903. Photo by Shepard. Minnesota Historical Society photo.

Yoerg met and married his future wife Elvina Seitzinger in 1851 in the home of Governor Alexander Ramsey. She was born in Pennsylvania and traveled with Ramsey to Minnesota in 1849 to serve as a governess for the Ramsey children.

In 1871 the brewery moved into three limestone buildings near the West Side caves at Ethel and Ohio Streets. The brew house was set directly into the cliffs, as the caves were a cool place to store beer. The site was the headquarters for the firm for the next 80 years.

Elvina Seitzinger Yoerg

Born in St. Paul, August 21, 1855, she attended St. Joseph's Academy. On June 28, 1877, she married John A. Seeger, giving birth to five daughters and two sons. She died, May 19, 1940.

William Seeger

William Seeger, about 1873. Minnesota Historical Society photo.

Born in Laumburg, Germany, on May 12, 1810, William Seeger studied at the College of Lübeck. He married Christine Wolf of Hamburg, Germany, on December 28, 1838. In 1852 he came to the United States, settling in Covington, Kentucky. In 1856 he settled in St. Peter, Minnesota, where he owned a brewery and later was engaged in the mercantile business in Le Sueur, Minnesota. He was active in bringing colonies of Russian and German immigrants to the state. During the Civil War he served in the Ninth Minnesota Regiment. Seeger served as State Treasurer in 1872–73 and as secretary of the state board of immigration.

John Augustus Seeger

Born in Covington, Kentucky, December 19, 1853, he was about two years old when the family moved to Minnesota. He completed his education in the public schools in St. Paul. In 1867, at the age of fourteen, he began an apprenticeship in the office of Leonard-Shiere, contractors, builders, and architects. In 1872 Leonard and Seeger formed a partnership, known as Leonard & Seeger Company, architects and builders. He was with the firm until 1877. Among their works of of St. Paul. He also aided the construction of the St. Paul Fire and Marine Insurance Company's building at the corner of Third and Jackson streets.

The association with Leonard proved to be an unhappy one for Seeger and his bride, Elvina Seitzinger Yoerg. When they returned from their wedding trip in 1877, they found that Leonard had abruptly left the firm and drained it of its cash. Desperate for operating funds to survive, Seeger turned to his new father-in-law, Anthony Yoerg, for assistance.

The founder of the first brewery in Minnesota in 1848, Yoerg was then riding a crest of business prosperity. In 1871 his firm had moved from its original location at Eagle and Washington Streets (today the site of the parking garage for St. Paul's RiverCentre) to the southwest corner of Ethel and Ohio Streets, across the Mississippi River from the Upper Landing and next to the West Side's legendary bluffs and caves. As the two families

John A. Seeger, about 1913. Civic-minded throughout his life, John Seeger was named "St. Paul's Leading Citizen" in 1929 by the Cosmopolitan Club. He had been a leader in the program of the United Improvement Council for the widening of streets, the building of Kellogg Boulevard, the construction of the city hall and county courthouse, and was chairman of the Auditorium Building Committee. He helped found the St. Paul Association of Commerce, St. Paul Athletic Club, the Empire National Bank, and the Anchor Casualty Company (he served as its president). He also constantly promoted the Community Chest and other charity drives. Minnesota Historical Society photo.

construction were the second Assumption Catholic Church (1874), which still stands at Seventh Street near St. Peter Street and the St. Paul Fire and Marine Insurance Building at Jackson and Third Streets. Seeger served alone as a contractor from 1877 to 1883, when he became the secretary of the Bohn Manufacturing Company, fabricating sashes, doors, and interior woodwork. In 1899 he advanced to president of the Bohn Corporation and remained so until 1905.

In 1897 Seeger established a plant for the manufacture of refrigerators and in 1902 formed the Seeger-Gallasch Refrigerator Company. He was made secretary-treasurer, and later president of the Seeger Refrigerator Company. In his business he witnessed and was instrumental in practically revolutionizing the manufacture of refrigerating works for residential and commercial purposes as well as for railroad freight and dining cars.

Seeger also served as president of the Anchor Casualty Insurance Company and as a director of the Northwestern Bank and of the Empire National Bank. For many years he was a member of the St. Paul Association of Commerce, the United Improvement Council, and the St. Paul Library Board.

Walter Gustave Seeger

Born in St. Paul, March 8, 1886, Walter Seeger attended the Van Buren grade school and was a graduate of Mechanic Arts High School. He attended the University of Minnesota for one year before joining his father's company located on East Seventh and Pine streets in 1907. A handful of men turned out each week's production of a few cabinets that were often delivered to the customer by a one-horse wagon hired for the job. Seeger once commented that "Fifty years ago when I first started peddling iceboxes for Seeger Refrigerator, the only things that were inseparable were the box and the ice." He soon was sent on the road to sell iceboxes and the first big break for the company came when large orders were secured for John Wanamaker in New York City and Marshall Field in Chicago. He served successively as secretary-treasurer (1908), vice president (1916), and on the death of his father, president of the company (1938). In 1950 he became chairman of the Seeger Refrigerator Company. Walter Seeger also served on the board of directors of the Great Northern Railway, the Northwest Bancorporation, the Empire National Bank, Northwestern Bell Telephone Company, Anchor Casualty Company, Northern Federal Savings and Loan Association, Gould National Batteries, and as a trustee of the Minnesota

continued on next page

lived near each other on Isabel Avenue near Ohio Street, they doubtless discussed Seeger's circumstances. Yoerg loaned his son-in-law the funds to maintain the company for the next eight years. In 1877 Seeger became an independent contractor and for approximately the next six years he built some of St. Paul's early buildings. The *1880–1881 St. Paul City Directory* listed John A. Seeger's business interests as "Lumber and Contractor and Builder, yard corner of St. Peter and Sixth, shop, corner of Eagle and Washington." Besides the financial assistance from his father-in-law, Seeger was using the site of the first Yoerg's brewery for his cabinet-making work.

Seeger joined the Bohn Manufacturing Company (a builder of sash, doors, blinds, and interior woodwork) as secretary in 1883, becoming a vice-president of the firm ten years later and its president in 1899. He oversaw the construction of the new Bohn factory at 835 Arcade Street in 1886 and

continued from previous page

Mutual Life Insurance Company. He was president of the St. Paul Baseball Club of the American Association league (1934–52). He married Nelle I. Luse on April 14, 1909.

Gustave R. Seeger

A St. Paul native, born January 16, 1889, he attended the Van Buren grade school and was a graduate of the Mechanic Arts High School. He attended the University of Minnesota and joined his father's company about 1908 serving first in sales and later as vice president of manufacturing operations and treasurer, posts that he held until his death in 1945. He married Martha T. R. Kieckhefer, of Milwaukee, Wisconsin, on August 30, 1913. They had three daughters and one son.

Gebhard Bohn

Born in Immenhausen, Hesse-Cassal, Germany, on May 11, 1854, Gebhard Bohn was educated in the local schools and at a polytechnic school, graduating in 1872. With his brother he migrated to Winona, Minnesota, where his brother started a sash and door business and Gebhard kept the books. He married Lena Nockin in 1875 and by 1880 was living in St. Paul where he became manager of a branch of his brother's company. In 1880 the business was located at Third and Jackson streets and moved the next year to Sixth and Wacouta streets. In 1887 the Bohn Manufacturing company located on Arcade Street at the site that became the home of the Seeger Refrigerator Company in 1910.

In the 1890s Bohn established with John Seeger the White Enamel Refrigerator Company, a subsidiary of the Bohn Manufacturing Company. A disagreement with other members of the Bohn family, however, prompted Gebhard Bohn to leave the parent company with the assets of the refrigerator operation and establish a new firm in the Market Building at Seventh and Wacouta streets. The company later located on Jackson Street between Fifth and Sixth streets and in 1902 to its twenty-three acre site at 1340 to 1400 University Avenue. He died July 23, 1924.

Gebhard Bohn, about 1905. Engraving by E.G. Williams and Brothers. Minnesota Historical Society collections.

Bohn Refrigerator Company

The company's origin in the 1890s is as the White Enamel Refrigerator Company, a subsidiary of the Bohn Manufacturing Company, located on Arcade Street. Its name was changed in 1918 to the Bohn Refrigerator Company. The company manufactured residential and commercial refrigeration products. In constant litigation with the Seeger Refrigerator Company over the use of patents between 1906 and 1935, it was never successful in its legal efforts. The company closed its doors and went out of business in 1935.

1887 at a cost of $55,000.[2] During this period Seeger began a long association with James J. Hill, manufacturing interior millwork for Hill's mansion under construction on Summit Avenue. This was a relationship that continued with the building of Louis W. Hill's adjacent house and the erection of the first buildings in Montana of what is today Glacier National Park.

Seeger's association with the Bohn Manufacturing Company proved to be difficult, discordant, and litigious. As early as 1896 the company experienced financial difficulties, a pattern that would recur with increasing complexity during the next eight years. In late December 1899, James J. Hill loaned the business and its officers, Gebhard Bohn, George W. Bohn, and John A. Seeger, $22,000, to meet a financial crisis.[3] Various banks in St. Paul and nearby country towns attempted to aid the company with loans on a short-term basis, but the cash transfusions proved useless.

During this period Bohn Manufacturing Company received a subcontract from the Butler Brothers Construction Company, the general contractors for the building of the new state Capitol designed by Cass Gilbert, to provide the interior woodwork. While the project was underway, the Bohn Manufacturing Company endured intense financial pressures and had to assign to the National German American Bank on September 15, 1903, its entire claim for payment of materials to be furnished for the new Capitol, valued at $65,842. In August 1904 the firm collapsed in bankruptcy and its operations closed.[4]

The earliest known photograph of the Seeger Siphon Refrigerator factory, 339-341-343 East Seventh at Pine Street, in 1904 or 1905. The two men in this photograph were the firm's first employees and both had worked for John A. Seeger as carpenters at the Bohn Manufacturing Company. John G. Fisher is at left, with Gustaf F. Steen on the right. Photograph courtesy of Mrs. Robert M. Reay

The exterior of the Seeger Refrigerator Company, 1902–1910, at 339, 341, and 343 East Seventh and Pine Streets. The original buildings are no longer standing and today this site is occupied by a gas station. Whirlpool Corporation photo.

As early as 1897 the Bohn Company had begun to build iceboxes as a sideline to its regular business, with Seeger in charge of the factory operations. For two years the company manufactured the iceboxes, whereupon Gebhard Bohn, the president of the firm, and his son Gebhard C. Bohn, severed connections with the business and established the White Enamel Refrigerator Company. The new firm bought the refrigerator stock and manufacturing equipment from the Bohn Manufacturing Company. Interestingly, in 1899 Adolph G. Gallasch joined the White Enamel Refrigerator Company as a one-third partner but left after little more than a year as the firm was suffering substantial losses.

In 1902, while serving as president of Bohn Company, Seeger formed a partnership with Gallasch in 1902 known as the Seeger-Gallasch Company.[5] The firm, located at 339–341–343 East Seventh at Pine Street, made cabinets as well as iceboxes and ice cream cabinets. By 1905 the firm was known as the Seeger Siphon Refrigerator Company. Three years later the corporation was renamed the Seeger Refrigerator Company and two of John A. Seeger's sons, Walter G. and Gustave R., joined him in operating the company. The Schuneman & Evans Department Store sold the young firm's products in St. Paul, as did the Boutell Brothers Furniture Store in Minneapolis.

At the same time, the Seeger firm also installed its patented "Siphon System" of refrigeration in railroad freight cars for several railroad companies including the Great Northern, the Northern Pacific, the Chicago, Milwaukee & St. Paul, the Chicago, Burlington & Quincy, the New York Central, and the Pennsylvania.

By 1906 Gebhard Bohn's firm, the White Enamel Refrigerator Company,

The cover of the 1896 catalog for the G.F. Quinn Refrigerator Company. On page 3 of this catalog is a diagram that was part of the patent suit brought against the Seeger Company. Quinn Refrigerator Catalogue in the author's collection.

had become the American Car and Foundry Company. The firm was manufacturing refrigerator cars for railroads and refrigeration units for dining cars based on the siphon system, as was Seeger. The interests of Bohn and Seeger were on a collision course since both firms claimed to be manufacturing refrigerators using the same cooling process. In 1906 Seeger-Gallasch initiated legal action against the American Car & Foundry Company to clarify its right to use the siphon cooling patent and process of the G. F. Quinn Refrigerator Company of Portland, Maine, on its products.[6]

A series of events stretching over nearly a decade describes the dispute.

ADOLPH G. GALLASCH JOHN A. SEEGER

Seeger Siphon Refrigerator Company

MANUFACTURERS OF

High Grade Dry Air Siphon White Enamel Lined Refrigerators

No. 10—Style W.

Style W, panel door	$28.00		
Style G, glass door	31.00		

DIMENSIONS IN INCHES.

	Width	Depth	Height
Outside	35½	20½	41
Provision Chamber	13½	14	28
Ice Chamber	11	14	22
Capacity of Ice Chamber			65 lbs.
Shipping weight			200 lbs

No. 25—Style W.

Style W, panel door	$37.00		
Style G, glass door	40.00		

DIMENSIONS IN INCHES.

	Width	Depth	Height
Outside	35½	20½	50
Provision chamber	13½	14	37
Ice chamber	11	14	31
Capacity of ice chamber			100 lbs.
Shipping weight			260 lbs.

SEEGER SIPHON SYSTEM

DIRECT ACTION.

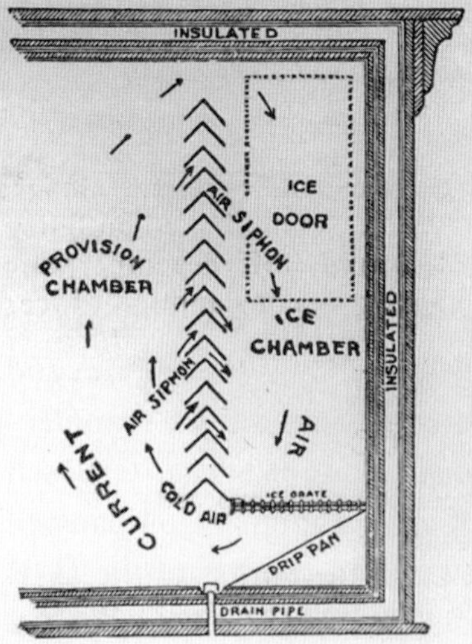

TO THE PUBLIC.

In presenting this catalogue showing a cut of our Siphon System and only a few of the various styles of Refrigerators manufactured by us, we wish to say that our Refrigerators are recognized by experts and all who have used them as the best at present upon the market.

We trust before purchasing you will make a thorough investigation of our system and claims. We guarantee the Refrigerator as represented.

We build and apply this system to every form of Refrigerators, from the ordinary House Refrigerator to those of the Dining, Buffet, and Refrigerator Cars, Creameries, Grocers, Florists, and Saloons.

Please call, or write us NOW for catalogue and prices. You certainly want the best—we have it and are ready to serve you.

Seeger Siphon Refrigerator Co.,
341-343 East Seventh Street, ST. PAUL, MINN.

ALSO

SCHUNEMAN & EVANS, St. Paul — BOUTELL BROS., Minneapolis

These two pages are from a 1905 catalog of refrigerators manufactured by the Seeger-Gallasch Company. The diagram on the bottom left of the page on the left shows the Seeger Original Siphon System for the refrigerated preservation of foods. This system was designed to keep air continuously in circulation throughout the interior of the icebox. Refrigerators such as these could be purchased directly from the company, or

Physicians praise our Refrigerators because they are constructed on sanitary lines and are the *Best Made.*

No. 30—Style W.

Style W, panel door........................ $40.00
Style G, glass door........................ 43.00

DIMENSIONS IN INCHES.

	Width	Depth	Height
Outside..................................	41¼	22	50
Provision chamber................	18	16	37
Ice chamber...........................	12½	15	31

Capacity of ice chamber........................... 120 lbs.
Shipping weight...................................... 315 lbs.

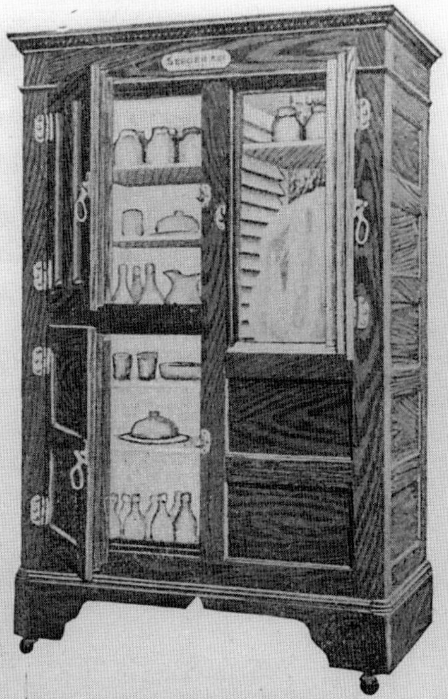

No. 38—Style W. W.

Style W. W., two panel doors....... $55.00
Style U.G., upper door glass, lower door panel.............................. 57.00
Style G. G., two glass doors......... 60.00

DIMENSIONS IN INCHES.

	Width	Depth	Height
Outside..................................	41½	22¾	66
Provision chamber................	18	16	53
Ice chamber...........................	12½	16	47

Capacity of ice chamber........................... 160 lbs.
Shipping weight...................................... 400 lbs.
Glass or panel doors as desired.

Important Points Necessary in Refrigeration

The Seeger Dry Air Siphon Enamel Lined Refrigerator provides for all of these:—

Fresh, pure and dry atmosphere;
Continuous circulation, which
Carries off and destroys all odors;
Absence of moisture and mold;

Sanitary and perfect drainage;
White enamel baked-on lining;
Low and uniform temperature;
Saving in consumption of ice.

Guarantee

Every *Seeger Siphon Refrigerator* is guaranteed as represented by us. If after a week's trial the refrigerator does not do what we claim for it, write us and if we cannot in every particular satisfy you we will take the refrigerator back, pay freight both ways, and refund your money. We ask you, isn't that fair? Write us.

from the Schuneman & Evans Department Store in St. Paul, or from Boutell Brothers Furniture Store in Minneapolis. Minnesota Historical Society photos.

An illustration of the Quinn Refrigerator Company's patented system of providing a refrigeration process. Quinn Refrigerator Catalogue in the author's collection.

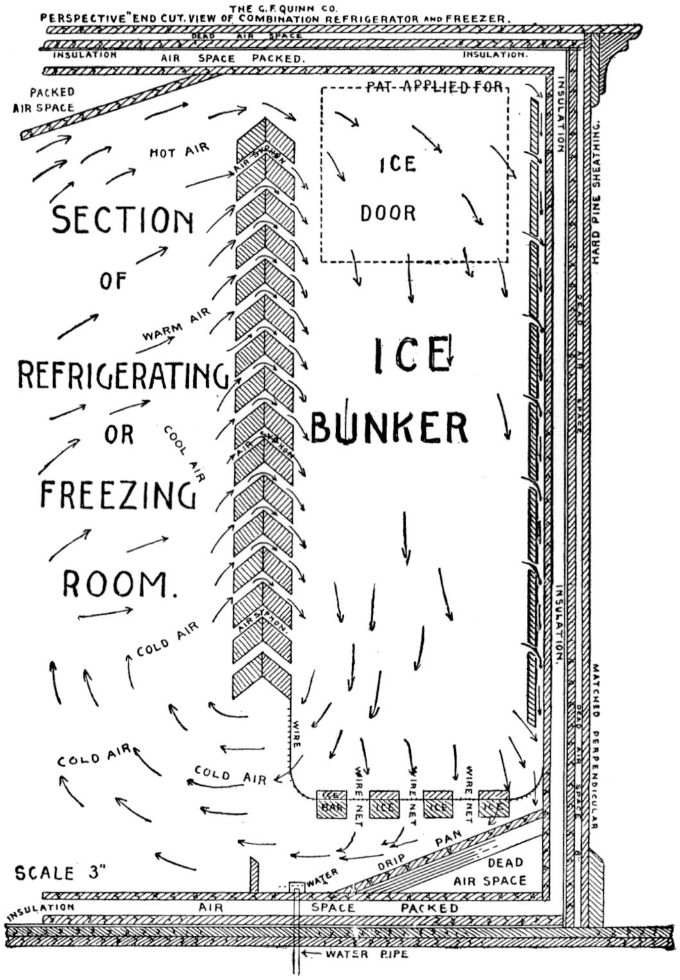

SECTIONAL VIEW OF REFRIGERATOR,

SHOWING CIRCULATION OF

Cold, Dry Air.

This system is protected in U. S. Patent Office, Washington, Great Britain, France and Germany. All infringements will be prosecuted.

The G. F. Quinn Refrigerator Company assigned the use of his patent on August 27, 1897, to Gebhard Bohn, George W. Bohn, John A. Seeger, and John H. Ames, the officers of the Bohn Manufacturing Company. In May 1898, for reasons that are not clear (quite possibly certain fees had not been paid to him), the Quinn Company assigned to Fred H. Johnson of Portland, Maine, all rights to the siphon system patent. The next year Johnson transferred the rights to a new entity, known again as the Quinn Refrigerator Company. In 1905 the Quinn Company transferred the well-traveled patent

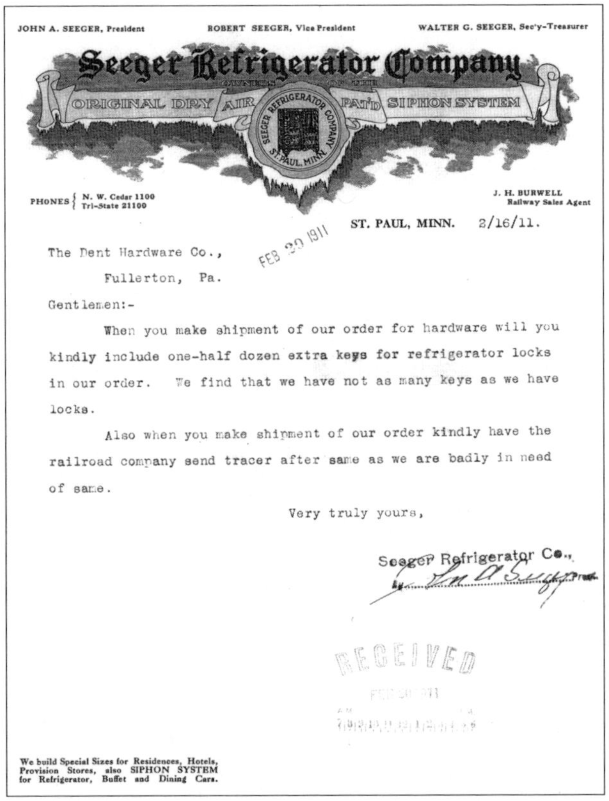

JOHN A. SEEGER, President ROBERT SEEGER, Vice President WALTER G. SEEGER, Sec'y-Treasurer

Seeger Refrigerator Company

ORIGINAL DRY AIR PAT'D SIPHON SYSTEM

PHONES { N. W. Cedar 1100 / Tri-State 21100

J. H. BURWELL Railway Sales Agent

ST. PAUL, MINN. 2/16/11.

FEB 20 1911

The Dent Hardware Co.,

Fullerton, Pa.

Gentlemen:-

When you make shipment of our order for hardware will you kindly include one-half dozen extra keys for refrigerator locks in our order. We find that we have not as many keys as we have locks.

Also when you make shipment of our order kindly have the railroad company send tracer after same as we are badly in need of same.

Very truly yours,

Seeger Refrigerator Co.,

RECEIVED

We build Special Sizes for Residences, Hotels, Provision Stores, also SIPHON SYSTEM for Refrigerator, Buffet and Dining Cars.

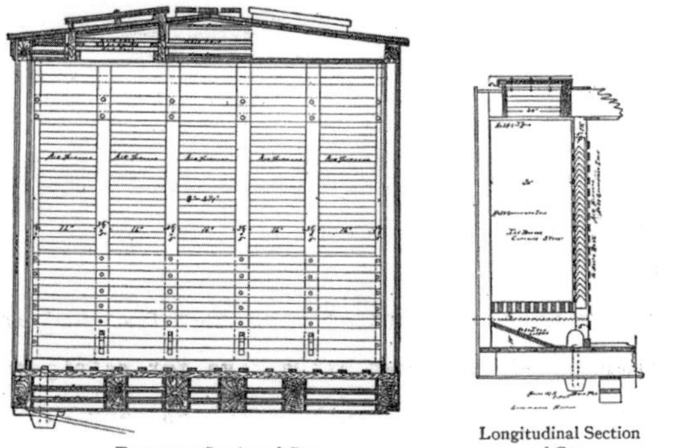

Transverse Section of Car. Longitudinal Section of Car.

A Few Railroads Who are Using Our System

New York Central & Hudson River R. R. Co.
New York, Chicago & St. Louis R. R. Co.
Cleveland, Cincinnati, Chicago & St. Louis Ry.
Lake Shore & Michigan Southern
Pittsburgh & Lake Erie
Frisco Lines
Chicago & Eastern Illinois R.R.Co.
Evansville & Terre Haute R.R. Co.
Chicago, New York & Boston Refrigerator Car Co.
Missouri Pacific R. R. Co.
Merchants' Despatch Transportation Co.
Illinois Central R. R. Co.
Great Northern R. R. Co.
Northern Pacific R. R. Co.
Barney Smith Car Co.
Pullman Car Co.
Chicago, Milwaukee & St. Paul R. R. Co.
Hicks Locomotive & Car Works
Chicago, Burlington & Quincy R. R. Co.
Pennsylvania R. R. Co.
Chicago, Rock Island & Pacific R. R. Co.
Wichita Falls & Northwestern R. R. Co.
And many others.

A 1911 letter from John A. Seeger to the dent Hardware Company in Fullerton, Pennsylvania. It shows an early version of the Seeger letterhead. On the reverse side of the letter, there is a blueprint of the Seeger refrigeration system installed in railroad freight cars. Letter in the author's collection.

for $8,000 to Adolph G. Gallasch, including "its entire right, title and interest in and to said Letter, Patent and to the invention or improvements shown and described therein and secured thereby." On September 29, 1906, Gallasch assigned to John A. Seeger all rights, title and interest in the patent. The Seeger Company owned the original patent on a dry air siphon system for railroad refrigerator and dining cars.

The Seeger application of the Quinn patent involved insertion of an open inverted "V" partition between the ice and storage chambers thereby allowing for easy, or siphoned, passage of cool or "refrigerated" air around perishable produce. It was this series of open spaces resembling siphons placed between the ice and refrigerated bunkers that distinguished the Quinn patent from earlier patents. The Quinn achievement was to introduce into refrigerators a new method of promoting the circulation of a confined body of air by the use of the open air spaces in a partition.

The legal case brought by Seeger against the White Enamel Refrigerator Company dragged on in U.S. District courts in New Jersey, New York, Minnesota, and before the Supreme Court for twelve years, and cost thousands of dollars. The Bohn defense enlisted the services of law firms on its behalf in St. Paul, Minneapolis, New York, Boston, Baltimore, and St. Louis. In St. Paul Bohn was represented by the firm of Davis, Kellogg and

These two pages show examples of commercial refrigerator cabinets that Seeger manufactured for use by hotels, hospitals, and others. From a Seeger catalog in the author's collection.

543
1905

No. 50—Style U. G.

South St. Paul, Minn., Aug. 8, 1905.
Seeger Siphon Refrigerator Co.,
St. Paul, Minn.
Gents—Enclosed please find check. I am more than pleased with the refrigerator.
Respectfully yours,
CHAS. SPILKER.

BUILT TO ORDER FOR JOS. GRUBER.

St. Paul, Minn., July 15, 1905.
Seeger Siphon Refrigerator Co.,
343 E. 7th St., St. Paul, Minn.
Gentlemen:—I take pleasure in saying that the special Refrigerator you built for me is in every way entirely satisfactory. It does not consume near as much ice as my old one, the temperature is lower, the circulation is perfect and the air always dry and cool. Your system is the best I ever saw and do not see how it could be improved upon.
Very truly yours,
261 W. 3d St., St. Paul, Minn. JOS. GRUBER.

No. 50—Style U. G.

No. 50—Style W. W., two panel doors...... $69.00
No. 50—Style U. G., upper door glass, lower door panel.................................. 72.00
No. 50—Style G. G., two glass doors........ 75.00

DIMENSIONS IN INCHES.

	Width	Depth	Height
Outside	49½	26¾	66
Provision chamber	23½	20½	53
Ice chamber	15	20½	47

Capacity of ice chamber..........................235 lbs.
Shipping weight...530 lbs.

WORKMANSHIP AND MATERIAL.

Our aim is to make the best refrigerator on the market—and we do. The panel work is made of the best kiln-dried hardwood. We could save nearly half of the expense by building the walls and linings cheaply, but that we will not do. We take pride and pleasure in producing a refrigerator of merit and efficiency. The provision chamber and shelves are all white enamel, thus always clean. The appearance of the refrigerator is inviting and at once inspires all persons with a desire to possess the **Seeger White Enamel Siphon System Refrigerator.** Value, not cheapness, should govern the construction of a refrigerator.

TESTIMONIALS.

GENTLEMEN: Auburndale, Mass.
The refrigerator you furnished Lasell Seminary proves all you claim for it. Consider it requires less ice and provides a colder atmosphere than any refrigerator I have ever seen. The circulation of cold, dry air is in constant attendance, preserving food indefinitely.
WOODLAND PARK HOTEL.

GENTLEMEN: Portland, Me.
I take great pleasure in saying a few words in praise of your refrigerator. I have used one for the past year, and am happy to say it has given perfect satisfaction in every respect; so much so that we are about to put in a much larger one. I find that it makes a great saving in the amount of ice used to keep it even at a freezing temperature, and can also say that food put in the Seeger refrigerator gathers no moisture, and therefore keeps much better; a point worthy of consideration in any refrigerator.
G. C. SHOREY,
Steward Maine Hospital.

GENTLEMEN: Boston, Mass.
Our house has your refrigerators in kitchen, store-room, and bar, and we are perfectly satisfied with them.
HOTEL SAVOY.

Butcher and Creamery Refrigerators

BUTCHER'S REFRIGERATOR.

BUTCHER BOX MADE TO ORDER.

TESTIMONIALS.

GENTLEMEN: Portland, Me.

After considerable investigation among some of our friends in the provision business, we decided that the Seeger Siphon Dry Air System was the best system of refrigeration we had ever seen. Therefore we had one installed in our shop, and are pleased to say that after a thorough trial we are more than satisfied, and cheerfully recommend it to dealers who want the best results obtainable in a refrigerator.

Yours,
CHASE & OWEN.

GENTLEMEN: Boston, Mass.

Replying to your inquiry as to how the refrigerator that you put into my house is working, I would say that it is giving perfect satisfaction. It is economical in the use of ice, always sweet and clean, and in fact does everything that you claim for it. You may refer intending purchasers to me at any time.

W. WALLACE WAUGH,
Manager "The Hotel."

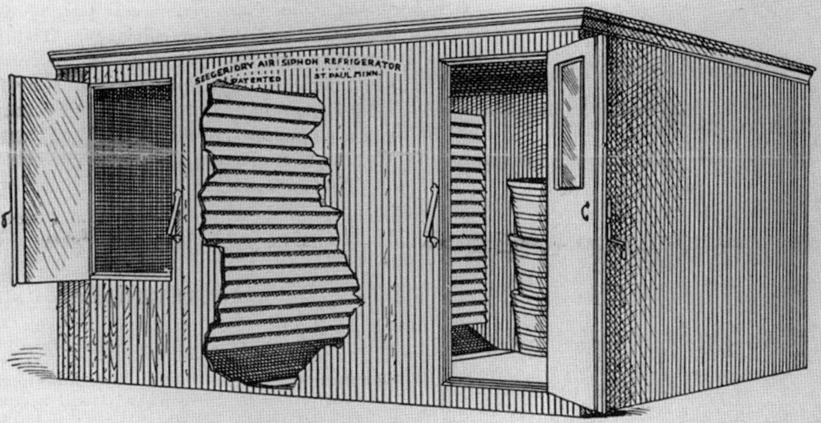

This cut shows our Dry Air Siphon Creamery Refrigerator, without question the best refrigerator on the market to-day, for the following reasons—

Low temperature from 36 to 42 degrees;
Atmosphere perfectly dry,
Direct circulation and air current
Absolutely free from moisture and mold,
The direct current prevents mold,
Also requires less ice.

We make the above claims and stand ready to guarantee all we have said. We make to order any size required. Our standard sizes hold from 80 to 210 tubs. Tell us about what you need and we will send you sketch and prices. Write us now. When in the city call at our office and factory, and we shall be glad to welcome you and explain to you why our Creamery refrigerators are superior to all others. It can be truly said, **it's the dairyman's friend.**

Send for Illustrated Catalogue and Price List

Seeger Siphon Refrigerator Company

341-343 East Seventh Street

St. Paul, Minn.

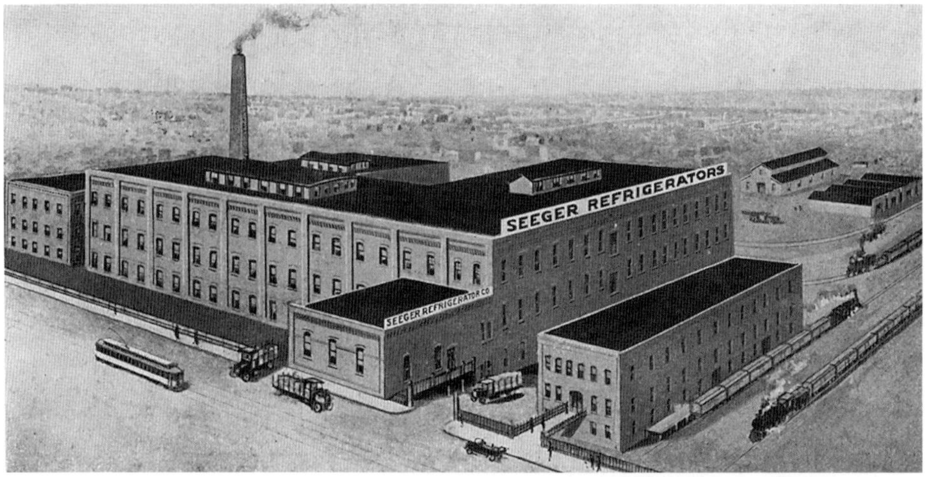

The Seeger Company factory on Arcade Street on St. Paul's East Side in the early 1930s. From a Seeger catalog in the author's collection.

Severance (now Briggs and Morgan), while Seeger-Gallasch was represented by Butler, Mitchell and Doherty, a predecessor of the twentieth-century firm of Doherty, Rumble and Butler. The case was heard in the U.S. District Court between July 22 and August 5, 1918. The court decided in favor of the Seeger Refrigerator Company, giving the firm a strong marketing impetus in an emerging household and commercial market. The victory of the Seeger firm in this case was not, however, the end of litigation between Bohn and Seeger as litigation continued in federal court over a variety of issues related to distinctive manufacturing operations until at least 1931. Within a few years the Bohn Company went out of business. For a complete genealogy of the Seeger business interests from the late nineteenth century to its merger with the Whirlpool Corporation and the appliance division of the RCA Corporation in the 1950s, see Figure 1.

From 1888 until its 1904 liquidation, the Bohn Manufacturing Company's plant was located at the southeast corner of Arcade Street and Wells Street. Ironically, and undoubtedly with a desire to rekindle and recover his nearly twenty-year association with the failed firm, John A. Seeger, during the difficult litigation against Gebhard Bohn and the American Car and Foundry Company, purchased the old Bohn facilities in 1910 from the St. Paul Title and Insurance Company. Out of the production and warehouse facilities of the old Bohn factory, a portion of the fourteen acres that would later comprise the Seeger manufacturing facility, Seeger would build his enterprise. A small woodworking department occupied the ground floor, with the cabinetmakers and assemblers on the second floor. The finishing and crating tasks were handled on the top floor. The Seeger Refrigerator Company plant had a long life, serving as a manufacturing facility until it was closed on December 28, 1984.[7] Today a few remnants of its extensive assembly lines and warehouses remain adjacent to the buildings of Seeger Square and reaching from Arcade Street on the west to Forest Street on the east of the site.

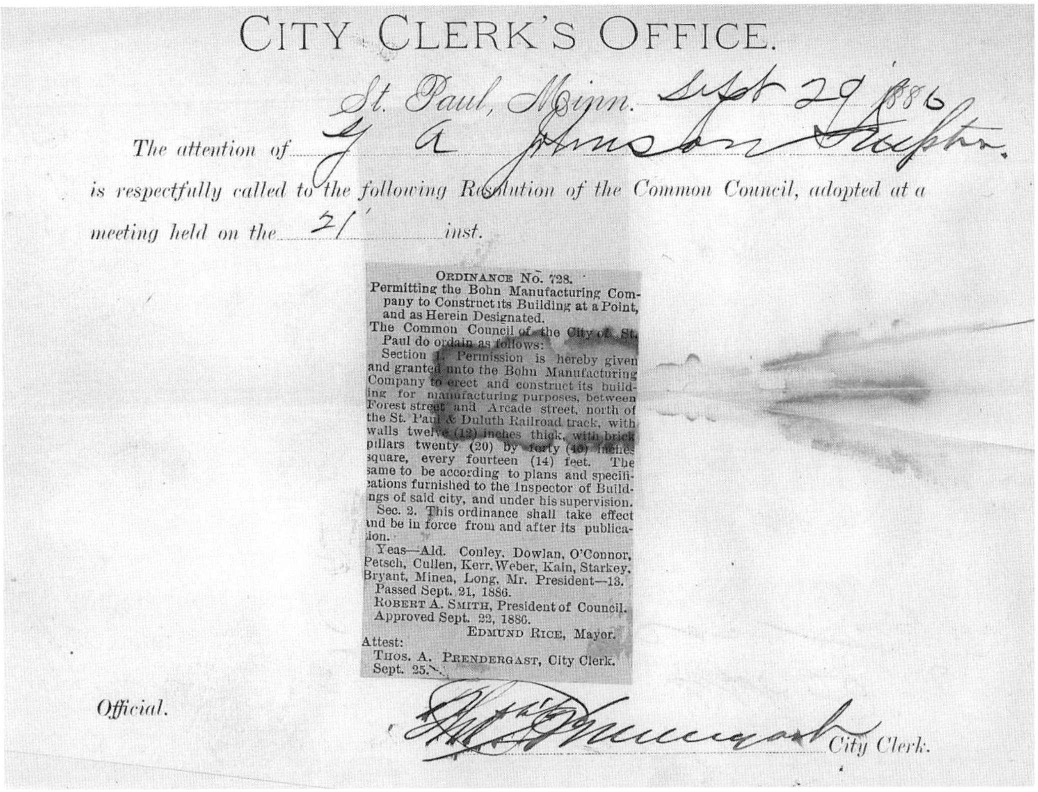

CITY CLERK'S OFFICE.

St. Paul, Minn. Sept 29 1886

The attention of J. A. Hinson [illegible]

is respectfully called to the following Resolution of the Common Council, adopted at a meeting held on the 21 *inst.*

ORDINANCE No. 728.
Permitting the Bohn Manufacturing Company to Construct its Building at a Point, and as Herein Designated.
The Common Council of the City of St. Paul do ordain as follows:
Section 1. Permission is hereby given and granted unto the Bohn Manufacturing Company to erect and construct its building for manufacturing purposes, between Forest street and Arcade street, north of the St. Paul & Duluth Railroad track, with walls twelve (12) inches thick, with brick pillars twenty (20) by forty (40) inches square, every fourteen (14) feet. The same to be according to plans and specifications furnished to the Inspector of Buildings of said city, and under his supervision.
Sec. 2. This ordinance shall take effect and be in force from and after its publication.
Yeas—Ald. Conley, Dowlan, O'Connor, Petsch, Cullen, Kerr, Weber, Kain, Starkey, Bryant, Minea, Long, Mr. President—13.
Passed Sept. 21, 1886.
ROBERT A. SMITH, President of Council.
Approved Sept. 22, 1886.
EDMUND RICE, Mayor.
Attest:
THOS. A. PRENDERGAST, City Clerk.
Sept. 25.

Official. Thos A Prendergast *City Clerk.*

In 1886 the St. Paul Common Council (today's City Council) granted approval for the Bohn Company to construct a manufacturing plant on the east side of Arcade and Wells Streets. St. Paul Building Permits Collection, Ramsey County Historical Society, St. Paul, Minn.

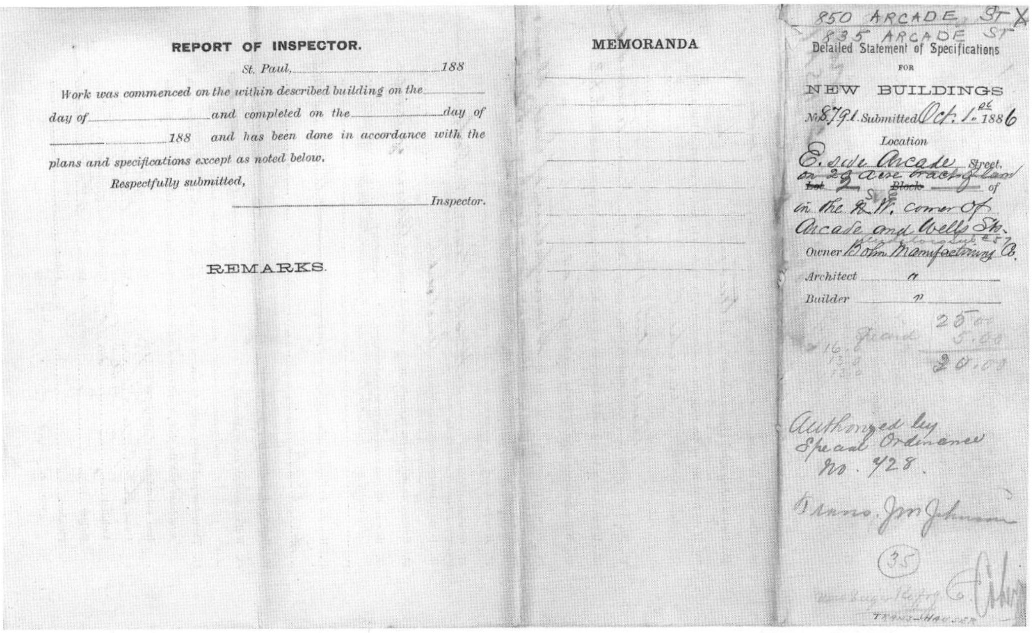

REPORT OF INSPECTOR.

St. Paul, ______ *188*

Work was commenced on the within described building on the ______ *day of* ______ *and completed on the* ______ *day of* ______ *188* *and has been done in accordance with the plans and specifications except as noted below.*

Respectfully submitted,

______ *Inspector.*

REMARKS.

MEMORANDA

850 ARCADE ST
835 ARCADE ST

Detailed Statement of Specifications FOR NEW BUILDINGS

No. 8791. *Submitted* Oct. 1st 1886

Location

E. side Arcade *Street,* on 22 acre tract ... of

in the N.W. Corner of Arcade and Wells Sts.

Owner Bohn Manufacturing Co.

Architect "

Builder "

25.00
5.00
20.00

Authorized by Special Ordinance No. 728.

Thos. Jm Johnson

(35)

The outside face of the 1886 building permit granted to the Bohn Manufacturing Company for their plant at 835–850 Arcade Street. St. Paul Building Permits Collection, Ramsey County Historical Society, St. Paul, Minn.

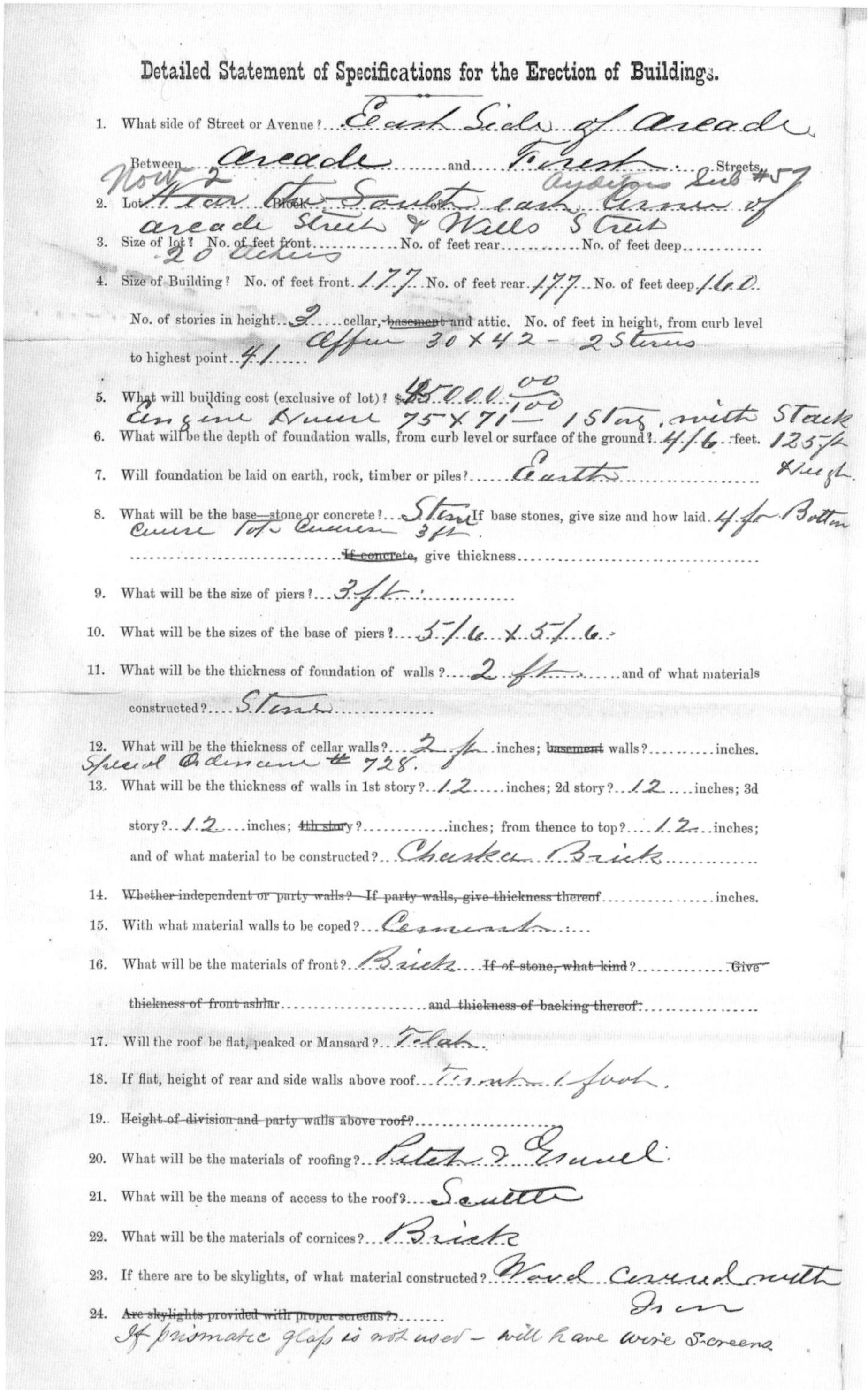

Detailed Statement of Specifications for the Erection of Buildings.

1. What side of Street or Avenue? East Side of Arcade

Between Arcade and Forest Streets. Aydejons Sub #57

2. Lot Now 7 Near the South east corner of Block Arcade Street & Wells Street

3. Size of lot? 20 acres No. of feet front............No. of feet rear............No. of feet deep............

4. Size of Building? No. of feet front 177 No. of feet rear 177 No. of feet deep 160

No. of stories in height 3 cellar, ~~basement~~ and attic. No. of feet in height, from curb level to highest point 41 Office 30 x 42 – 2 Stories

5. What will building cost (exclusive of lot)? $45000 00/100 Engine House 75 x 71 – 1 Story, with Stack

6. What will be the depth of foundation walls, from curb level or surface of the ground? 4/6 feet. 125 ft High

7. Will foundation be laid on earth, rock, timber or piles? Earth

8. What will be the base—stone or concrete? Stone If base stones, give size and how laid. 4 ft Bottom Course Top Course 3 ft.

...... ~~If concrete~~, give thickness......

9. What will be the size of piers? 3 ft.

10. What will be the sizes of the base of piers? 5/6 x 5/6

11. What will be the thickness of foundation of walls? 2 ft and of what materials constructed? Stone

12. What will be the thickness of cellar walls? 2 ft inches; ~~basement~~ walls?..........inches.

Special Ordinance #728.

13. What will be the thickness of walls in 1st story? 12 inches; 2d story? 12 inches; 3d story? 12 inches; ~~4th story~~?..........inches; from thence to top? 12 inches; and of what material to be constructed? Chaska Brick

14. ~~Whether independent or party walls? If party walls, give thickness thereof~~..........inches.

15. With what material walls to be coped? Cement

16. What will be the materials of front? Brick ~~If of stone, what kind?~~.......... ~~Give thickness of front ashlar~~.......... ~~and thickness of backing thereof?~~..........

17. Will the roof be flat, peaked or Mansard? Flat

18. If flat, height of rear and side walls above roof. Front 1 foot.

19. ~~Height of division and party walls above roof?~~..........

20. What will be the materials of roofing? Pitch & Gravel

21. What will be the means of access to the roof? Scuttle

22. What will be the materials of cornices? Brick

23. If there are to be skylights, of what material constructed? Wood covered with Iron

24. ~~Are skylights provided with proper screens?~~......

If prismatic glass is not used – will have wire Screens

This page and the one following reproduce the details of the plan for the Bohn plant on Arcade Street. The walls of the factory were to be twelve inches thick with brick pillars twenty by forty inches square and placed every fourteen feet. The building was to be three stories high, 177 feet in width, and 160 feet deep. Its estimated cost was $45,000. The document is signed by John A. Seeger for the Bohn Company. Seeger later purchased the site in 1910 and it was occupied until December 1984. St. Paul Building Permits Collection, Ramsey County Historical Society, St. Paul, Minn.

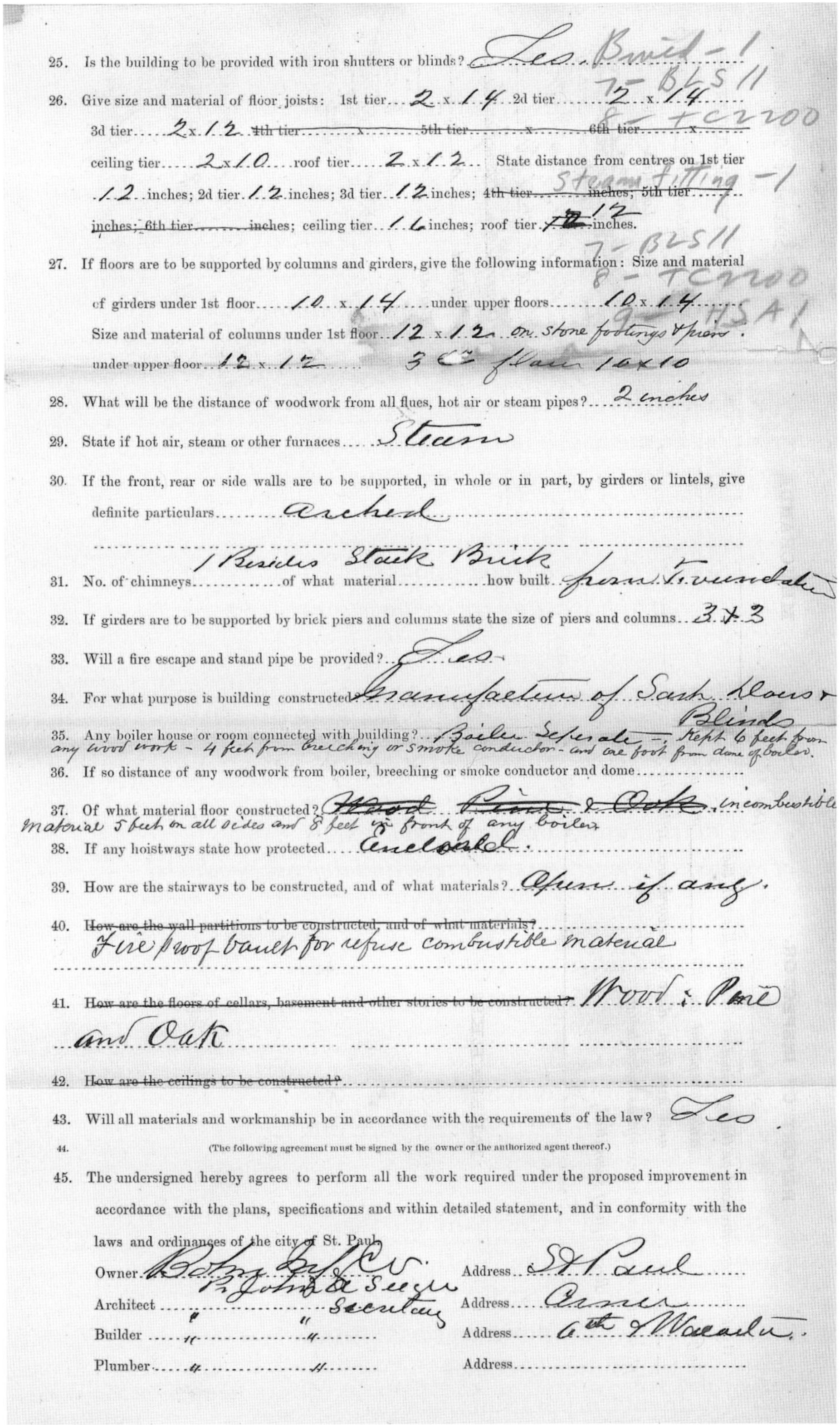

25. Is the building to be provided with iron shutters or blinds? Yes. Brick-1

7-BLS11

26. Give size and material of floor joists: 1st tier 2 x 14 2d tier 2 x 14

3d tier 2 x 12 ~~4th tier x 5th tier x 6th tier x~~ 8-TC2200

ceiling tier 2 x 10 roof tier 2 x 12 State distance from centres on 1st tier 12 inches; 2d tier 12 inches; 3d tier 12 inches; ~~4th tier inches; 5th tier inches; 6th tier inches~~; ceiling tier 16 inches; roof tier 12 inches. Steam fitting -1

7-BLS11

27. If floors are to be supported by columns and girders, give the following information: Size and material 8-TC2200

of girders under 1st floor 10 x 14 under upper floors 10 x 14

Size and material of columns under 1st floor 12 x 12 on stone footings & piers. 9-HSA1

under upper floor 12 x 12 3rd floor 10 x 10

28. What will be the distance of woodwork from all flues, hot air or steam pipes? 2 inches

29. State if hot air, steam or other furnaces Steam

30. If the front, rear or side walls are to be supported, in whole or in part, by girders or lintels, give definite particulars arched

31. No. of chimneys 1 Besides Stack of what material Brick how built from Foundation

32. If girders are to be supported by brick piers and columns state the size of piers and columns 3 + 3

33. Will a fire escape and stand pipe be provided? Yes.

34. For what purpose is building constructed Manufacture of Sash Doors & Blinds

35. Any boiler house or room connected with building? Boiler Seperate — Kept 6 feet from any wood work — 4 feet from breeching or smoke conductor — and one foot from dome of boiler.

36. If so distance of any woodwork from boiler, breeching or smoke conductor and dome

37. Of what material floor constructed? ~~Wood Pine & Oak~~ incombustible material 5 feet on all sides and 8 feet in front of any boiler

38. If any hoistways state how protected Enclosed.

39. How are the stairways to be constructed, and of what materials? Open if any.

40. ~~How are the wall partitions to be constructed, and of what materials?~~

Fire proof vault for refuse combustible material

41. ~~How are the floors of cellars, basement and other stories to be constructed?~~ Wood: Pine and Oak

42. ~~How are the ceilings to be constructed?~~

43. Will all materials and workmanship be in accordance with the requirements of the law? Yes.

44. (The following agreement must be signed by the owner or the authorized agent thereof.)

45. The undersigned hereby agrees to perform all the work required under the proposed improvement in accordance with the plans, specifications and within detailed statement, and in conformity with the laws and ordinances of the city of St. Paul.

Owner Bohn Mfg Co. Pr John A Seeger Secretary Address St Paul

Architect " " Address Owner

Builder " " Address 6th & Wacouta.

Plumber " " Address

Figure 1

A Genealogy of Seeger Business Interests

Leonard (1858–1872)
Leonard & Seeger (1872–1877)

↓

Seeger (1877–1883)
[engaged in lumber supply, contracting, and building]

↓

Bohn Manufacturing Company (1883–1904)
[manufacturer of sash, doors & blinds, packing boxes & shooks and supplier of kiln-dried hardwood lumber]

↓

Seeger-Gallasch (1902–1905)
[manufacturer of iceboxes and ice cream cabinets]

↓

Seeger Siphon Refrigerator (1905–1908)
[manufacturer of iceboxes and ice cream cabinets]

↓

Seeger Refrigerator Company (1908–1945)
[manufacturer of iceboxes, refrigerators, freezers, and defense products]

↓

Seeger-Sunbeam Corporation (1945–1946)
[manufacturer of refrigerators and freezers]

↓

Seeger Refrigerator Company (1946–1955)
[manufacturer of refrigerators, freezers, humidifiers, and icemakers]

↓

Seeger Refrigerator Company and the
RCA Appliance Division and the Whirlpool Corporation
merge (1955) to form
Whirlpool-Seeger Corporation (1955–1958)
which then becomes
Whirlpool Corporation (1958–)

2

From Arcade Street to Main Street: Iceboxes and Refrigerators

AT THE NEW ARCADE STREET PLANT in 1910, four Seeger family members, John A., Walter G., Gustave R., and John A.'s son-in-law, Charles A. Muessel, laid the foundation for an industrial concern that was destined to become one of the leading manufacturers of its field. With increased facilities, the company began to expand and reach markets farther afield. Walter Seeger and Gus Seeger became the sales force traveling about the country calling on department stores, hardware stores, and other similar outlets taking orders. From 1909 until about 1930, Charles Muessel launched and guided the firm's strategic national advertising campaign with company catalogs of its products and magazine advertisements in such publications as *Good Housekeeping*, the *Saturday Evening Post*, *Country Life*, *House and Gardens*, and *The National Geographic*. He also established and positioned branch sales offices in major cities from Boston, Massachusetts, to Los Angeles, California.

In these early days with limited orders and manufacturing entirely from wood and sheet steel, the largest percentage of the employees were carpenters and cabinetmakers and the business because of its nature was seasonable, lasting from March until November. Slowly and steadily the company grew until shortly Gustave R. Seeger left the Sales Department and devoted his entire time to the manufacturing end of the business while Walter G. Seeger and Charles A. Muessel continued in sales and John A. Seeger concentrated his attention on general management. The company grew in a steady manner under John A. Seeger's leadership, especially after his son Walter had traveled to Chicago, Philadelphia, and New York seeking orders for the firm. John A. Seeger was concerned about the seasonal nature of sales that experienced a heavy demand for its products during the spring, summer, and fall months, but saw a decline in orders during the winter season. To keep his carpenters busy year-round, Seeger sought long-term orders from key department stores in the East and Middle West. In New

The Major Players in the Seeger Story

Nelle Ienza Luse

Born in Carroll, Iowa, April 23, 1887, Nelle Luse was the daughter of Jesse Franklin Luse and Ida May Rice. They moved to St. Paul about 1904, where her father established the office of the Luse Land Development Company for the sale of more than four million acres in western Saskatchewan and two million acres in eastern Oregon. She attended St. Mary's Hall in Faribault for at least two years before attending and graduating from Oak Hall in St. Paul. She gave birth to three daughters. She died on August, 28, 1955.

Martha Kieckhefer Seeger

A native of Milwaukee, Wisconsin, Martha Kieckhefer was born February 4, 1889. Her family had arrived in the United States from Germany about 1850. In 1890 Martha's father, William, purchased the Enterprise Box and Lumber Company in Milwaukee. The name was then changed to the Kieckhefer Box Company. In 1957 the company was merged with the Weyerhauser Timber Company of Tacoma, Washington. Martha married Gustave Seeger on August 30, 1913. She died July 9, 1959.

Charles A. Muessel

Born in St. Paul, August 14, 1881, Charles Muessel married Bertha H. Seeger, daughter of John A. Seeger and Elvina Yoerg Seeger, on September 25, 1907. Some time prior to 1909, he joined the sales staff of the Seeger Company and was responsible for implementing an impressive strategic sales and marketing program for the young refrigerator manufacturer. He brilliantly launched a national advertising campaign for Seeger products in major publications and produced on a nearly annual basis a series of comprehensive catalogs of the company's products. Charles Muessel served as president of the Sales subsidiary of Seeger until at least 1930. In the 1920s he headed Seeger Sales offices in Boston and New York and was responsible for the installation of refrigeration equipment at many schools, colleges, universities, and hospitals in the region. In time he became a vice president of Seeger and served until his retirement in 1954. Muessel's efforts in many ways led and sustained Seeger's position in the appliance industry.

John Seeger Holl

The son of Anna V. Seeger and Walter C. Holl, John S. Holl was born in St. Paul, on December 1, 1905. His mother died in a tragic kitchen fire two years later and he was raised in the home of his grandparents, John and Elvina Seeger. He attended Carleton College in Northfield and joined Seeger in 1925. Rising through various company positions to serve as president of the firm in 1952, John Holl also was as a member of the board of directors for the Seeger Company and of the Whirlpool-Seeger and Whirlpool Corporation until 1970. He married Delores E. Ryder in 1958. Holl served on the board of directors of the First National Bank, the Children's Hospital, and as a trustee of Macalester College.

Louis W. Korfhage

Born in St. Paul, May 10, 1884, Louis Korfhage married Johanna E. Seeger, a daughter of John and Elvina Seeger, on November 8, 1911. A practicing dentist, he served for many years as a member of the Seeger board of directors.

Robert S. Aherns

Born in Cincinnati, Ohio, on June 22, 1890, his middle name was Seeger although he was not related to the St. Paul family. According to his daughter's recollection, he was named after a friend of the Aherns family named Seeger who lived in Cincinnati who may have been related to the St. Paul family. He arrived in St. Paul about 1919 and searching for a job, he called on John A. Seeger who invited him to the house for a Sunday night dinner. The meal turned out to be a propitious occasion as he met Freida E. Seeger, the youngest daughter of John and Elvina Seeger, and was offered a job at the Seeger factory. Freida and Robert were married September 18, 1920. While working at Seeger, Ahrens also attended the Dunwoody Institute in Minneapolis earning his qualifications as an engineer. Throughout his career he worked in the engineering department at Seeger and later served as a professor at the Dunwoody Institute. Robert Ahrens rose to become the chief engineer at the Seeger Company, a post that he held until his untimely accidental death in 1944.

Good Housekeeping Institute

Conducted by Good Housekeeping Magazine

This Certifies, that ... manufactured by ... has been tested by Good Housekeeping Institute, conducted by Good Housekeeping Magazine, and said article is hereby awarded the official approval of the Institute.

Tested and Approved by Good Housekeeping Institute

1917-18

The Good Housekeeping *Seal of Approval was first awarded to the Seeger refrigerator in 1913 by the Good Housekeeping Institute after extensive testing. The award certificate is on the top and the actual seals are below. From a Seeger catalog in the author's collection.*

York City he secured a commitment from the John Wanamaker Department Store to manufacture its iceboxes, while soon afterward Marshall Field in Chicago followed suit. Not much later the company supplied Seeger iceboxes to Sears, Roebuck and Company and to Montgomery Ward for their customers.

As early as 1911 Seeger began to manufacture refrigeration units for the U.S. Army and Navy for installation at various sites in the Panama Canal Zone. A few years later following the establishment of the Emergency Fleet Corporation in late 1916 as part of the federal government's initial effort to ready the nation for possible involvement in World War I, the company supplied refrigerators to merchant vessels that this public agency built, purchased, leased, or requisitioned for the shipment of food supplies. Similarly the U.S. Navy purchased Seeger refrigerators for many of its ships and both

The 1920 Seeger icebox with its oak exterior. The characteristic Seeger label with its script lettering is at the center of the top of the front panel. Whirlpool Corporation photo.

the army and the navy installed Seeger appliances at several of their large bases.[1]

Beginning in 1913 the Good Housekeeping Institute, which was conducted by *Good Housekeeping Magazine*, exhaustively tested the Seeger Refrigerator. The Institute substantiated the statement of the Seeger Company regarding the construction of its refrigerator and the appliance's efficiency, thereby granting the Institute's diploma and implicit endorsement to the Seeger appliance annually.

Between 1902 and 1930 catalogs of Seeger products illustrated the wide range of appliances manufactured for refrigeration. During these years, the company's basic line of appliances consisted of residential and commercial iceboxes and display cases, walk-in refrigerators for meat markets, and the installation of refrigeration systems in railroad cars. At the same time the company implemented the latest technological improvements in the con-

Seeger Women Investors

During the first twenty years of Seeger Company operations, the wives of Walter and Gustave Seeger made nearly equal investments in the company. Both women's assets were likely to have been transferred to them by their fathers.

Nellie Luse Seeger. Photo courtesy of John S. Reay.

Walter's wife, Nellie Luse, probably made her investment in either late 1916 or early 1917, after the sale of her father's company, the Luse Land Development Company, following his death in late 1916. For more than two decades Jesse Franklin Luse's St. Paul-based company had developed vast land holdings in Western Saskatchewan, Eastern Oregon, and Latin America. It is possible that proceeds from the sale of the land company or her inheritance was the basis of her purchase of common stock in the Seeger Company.

In 1916 and again in 1922 Seeger was expanding its manufacturing facilities to produce cabinets for General Electric, Kelvinator, Copeland, and other customers. Gustave's wife, Martha Kieckhefer, a Milwaukee native and daughter of the owner of the Kieckhefer Box Company, obtained shares of stock after a 1923 investment.* Her financial commitment to the company came at a time of further expansion of Seeger's Arcade Street plant.

Martha K. Seeger. Photo courtesy of Walter Andrews.

The exact circumstances that prompted these Seeger women to make their investments in the company remain unclear as the financial records of the corporation were systematically destroyed when the St. Paul factory was closed in December 1984. The limited number of records that has survived indicates, however, that their transactions definitely occurred during periods of expansion of the company's physical plant and production facilities.

*Seeger Refrigerator Company Shareholders Minutes, May 21, 28, 29; June 10; October 3, 1923.

struction and insulation of iceboxes. From a team of two cabinetmakers skillfully crafting iceboxes and ice cream chests in 1902, the company grew to employ more than 200 men by the early 1920s manufacturing 2,600 iceboxes and refrigerator cabinets a year. In 1935 the Seeger Company produced 275,000 units, and by 1941 it employed more than 2,500 persons and manufactured more than 300,000 cabinets yearly.

The first major expansion of the physical plant and production facilities occurred at Segeer between 1914 and 1919. The resulting change to the plant as a whole can be traced in the development of the more easily segregated porcelain enameling department. In the early days, as porcelain enamel came into popularity the sectional and exterior parts were purchased. Responding to the requests of customers Seeger in 1914 began to manufacture a line of porcelain clad exterior refrigerators and by 1919 the company established one of the first enameling plants in the industry.[2] The principal equipment in this plant consisted of a single, box-type furnace, two spray booths, a drier, and a few small wooden tanks for pickling or cleaning the parts to be processed. Each succeeding year additional equipment and supplemental furnaces were added until capacity increased to such a proportion in 1924 that the walls of the small plant fairly bulged.

About 1910 or 1912 Seeger began an association with the General Electric Corporation to develop a refrigerator that would carry the "G.E." label.

A typical page from a Seeger catalog. From a Seeger catalog in the author's collection.

Seeger Siphon Refrigerators

No. 20—Style W.

Style W, panel Door. Style G, glass door.

No. 20—Elastic enamel lined. List price $58.00
No. 200—Porcelain enamel lined. List price 69.00

For Glass Door, add $2.50 to net price.

Dimensions in Inches:

	Width	Depth	Height
Outside	36½	21½	45
Provision Chamber	16⅛	15¼	31¼
Small Provision Chamber	14⅞	15¼	9½
Ice Chamber	12¼	15¼	19¾

Capacity of Ice Chamber .... 100 lbs.
Shipping Weight .... 290 lbs.

White Enameling

The enameling is done under a certain process of our own, and baked in an oven built for the purpose, which is an assurance to buyers that they get the best. We are careful in selecting the galvanized steel and all materials that are used in the construction of our refrigerators. The shelves are made of woven wire heavily tinned also of galvanized steel white enameled.

Prices subject to liberal discount.

Page Seven

Seeger

ORIGINAL SIPHON SYSTEM

THE SEEGER ORIGINAL SIPHON SYSTEM OF CIRCULATION

A MAJOR FEATURE OF EFFICIENCY in the Seeger Refrigerator—a remarkable food, ice and power conserving device (whether ice or electrical refrigeration is used) is the Seeger System of Air Circulation—The Original Siphon System. It is installed only in Seeger Refrigerators and accomplishes the successful preservation of foods and the necessary low temperature with a minimum consumption of ice or electricity.

The Seeger Original Siphon System, briefly, continuously keeps in circulation, throughout the interior, a vigorous current of air that is dry and clean, and keeps the refrigerator's every nook and corner pure and sweet at all times.

The Siphons form a partition between the ice or coil chamber and the food chambers. The cold air—heavier in the ice or coil chamber than in the food compartments—continuously sinks to and through the grate beneath the ice block or coils, then to the slanting deflector plate that is seen beneath the grate. Next the deflector plate projects the air into the food chambers. In the food chambers the air expands, and in so doing consumes any and all heat atoms that are existent there, and picks up all odors, moisture and impurities.

Finally, the Siphons draw the air back into the ice or coil chamber where all the odors and impurities that have been gathered up are condensed and drained off with the water from the melting ice or coils. So long as any ice remains the circulation continues.

Seeger Original Siphon System

SUITABLE FOR ELECTRICAL REFRIGERATION WITHOUT CHANGE

[Three]

A diagram showing how the Seeger Original Siphon System worked in an icebox. From a Seeger catalog in the author's collection.

The earliest test models consisted of an oak Seeger cabinet with a G.E. Monitor Top compressor mounted on the top of the cabinet.[3] In 1916 the Seeger Refrigerator Company took a step that would change the scale of the company's structure. General Electric awarded Seeger a contract to construct the porcelain cabinets for GE's new line of monitor-top electric refrigerators. During these early days when porcelain enamel cabinet finishes were coming into popularity, Seeger purchased the sectional and exterior parts that it needed for its appliances. As will be seen later in this chapter, this manufacturing innovation had wide-ranging consequences for the future growth of the Seeger Company.

A brief outline of the company's sales operations emerges from the pages of a surviving minute book of the board of directors. The 1916 contract to manufacture refrigerator cabinets for the General Electric Corporation launched Seeger on a new era for production and sales. As late as 1928 General Electric maintained its association with Seeger, contracting for delivery of $1,000,000 worth of refrigerator cabinets, the equivalent of $10,664,242 in 2005 inflation adjusted dollars, for shipment to its factory

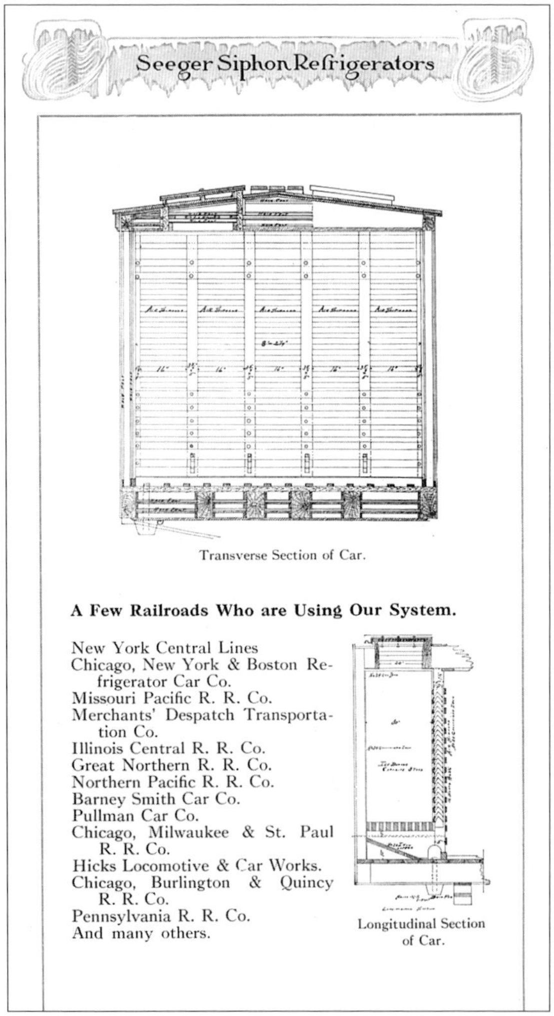
Seeger Siphon Refrigerators

Transverse Section of Car.

A Few Railroads Who are Using Our System.

New York Central Lines
Chicago, New York & Boston Refrigerator Car Co.
Missouri Pacific R. R. Co.
Merchants' Despatch Transportation Co.
Illinois Central R. R. Co.
Great Northern R. R. Co.
Northern Pacific R. R. Co.
Barney Smith Car Co.
Pullman Car Co.
Chicago, Milwaukee & St. Paul R. R. Co.
Hicks Locomotive & Car Works.
Chicago, Burlington & Quincy R. R. Co.
Pennsylvania R. R. Co.
And many others.

Longitudinal Section of Car.

This photo and the adjacent diagram are from a 1911 Seeger catalog. They show how the Seeger Siphon System functioned in refrigerated railroad cars. From a Seeger catalog in the author's collection.

in Covington, Kentucky for the installation of electric compressors.[4]

On February 16, 1916, the company established the Seeger Sales Corporation for the purpose of selling all of the products manufactured by the Seeger Refrigerator Company. Under the leadership of Charles A. Muessel, the Sales Corporation flourished until 1931 when changed circumstances prompted Seeger to terminate the Sales Corporation. To promote the manufacturing and sales of its refrigerators, the company's board of directors in October 1926, approved an annual budget of $150,000 for the printing of catalogues, circulars, printed matter, and national advertising of its products.[5] A year later, $200,00 was allotted for the purpose.[6] By 1928, perhaps recognizing the recently forged association with Sears, Roebuck and Company, the budget for promotional purposes was drastically reduced to $75,000, while in 1929 it increased slightly to $96,000.[7] The new relationship with Sears, Roebuck and Company to manufacture the *Coldspot* line of refrigerators caused

Seeger to significantly reduce its production of private-label appliances for other companies.

The wooden icebox was the staple of the company from its founding until about 1927. But times were quickly changing, particularly with the spread of electricity for home and commercial use. Early in the 1920s the appearance of the electric refrigerator changed the icebox industry. The best cabinets of this time were internally and externally covered with sheet porcelain and the outer edges protected with heavy, nickel-plated trim. The public demand for the new product was immediately positive and strong. A number of firms doing business on a large scale were caught unprepared for the technological and electrical advances in the industry. With foresight, imagination, good business sense, and two uncontrollable factors, excellent timing and good fortune, the Seeger Company rolled up its sleeves and went into action immediately.

The General Electric "Monitor Top" electric refrigerator with the cabinet made by the Seeger Company. The Seeger label is on the front panel at the top center. From a General Electric catalog in the author's collection.

About 1926 Seeger ceased manufacturing the wooden icebox and shifted to production of their own cabinets for ice, electrical, and mechanical refrigeration and as subcontractors for other companies selling the appliances under their own trade name. Again the company carefully noted further improvements in the construction, insulation, and design of the all porcelain appliances.[8] By 1930 Seeger offered such additional features for De Luxe residential models as an interior electric light and a "Chiltray" for the storage of particular foods and meats. The company also manufactured for home use sixteen different styles of refrigerators that could be used for ice, electrical, or mechanical refrigeration.[9] Seeger also offered an extensive line of thirty-two innovative porcelain commercial cabinets for storage and display for use by vendors, hospitals, and colleges.[10]

During the 1920s Seeger supplied cabinets to such manufacturers of electrical refrigerator compressors as General Electric, Kelvinator, Fairbanks Morse, Norge, Copeland, Servel, and over the years to more than twenty-five other manufacturers that applied their private labels to the finished product. After reaching an agreement with General Robert E. Wood, president of Sears, Roebuck and Company in 1928, Seeger Refrigerator Company joined with the Sunbeam Electric Manufacturing Company of Evansville in Indiana, to develop and manufacture the *Coldspot* refrigerator.

Seeger and the Murphy Trucking Company

For at least sixty years, and perhaps longer, the Seeger Refrigerator Company located on Arcade Street in St. Paul, and the Murphy Trucking Company maintained a continuing business relationship. The origins of the association is lost in the affairs of both companies during the 1920s and possibly as early as 1916. Whenever the mutually sustaining link started, it lasted until the 1970s. The Murphy Trucking Company was a family-owned business that began in St. Paul in 1904. In 1923 the company built a second warehouse in Minneapolis and was then known as the Murphy Transfer & Storage Company. Its president was an energetic Irish-American named Edward L. Murphy Sr., who had started his business with two horses and a wagon.

When Seeger began manufacturing the cabinets for Sears, Roebuck and Company's first electrical refrigerator—the *Coldspot*—in 1930, the Murphy firm transported the appliances from St. Paul to Evansville, Indiana, where the Sunbeam Electric Manufacturing Company installed its electric compressors in the refrigerator cabinets that Seeger had made for Sears.

It seems likely that Murphy also transported Seeger cabinets during the 1920s to several other companies that installed electrical units and placed their own brand names on the finished refrigerators. More than thirty companies used Seeger cabinets including Copeland, Kelvinator, Fairbanks-Morse, Norge, Westinghouse, and Amana. In 1916 Seeger began manufacturing the cabinets for General Electric Company's first refrigerator and Murphy may have transported the cabinets over the road from St. Paul to the G.E. assembly plant in Fort Wayne, Indiana.

During the several expansions of the production facilities at the Arcade Street plant from the 1920s to the 1950s, Seeger called on the Murphy Company for additional services. On numerous occasions Murphy supplied the industrial rigging and technical expertise for the installation of the massive industrial equipment and presses to be installed in the Seeger plant. The heavy mechanical machinery was used to stamp-out tools, parts, and cabinets for the manufacturing and assembly of Seeger's broad range of domestic and commercial refrigeration products.

As Murphy maintained extensive warehouse facilities in the St. Paul area and Seeger did not, the Seeger management contracted for Murphy trucks to transport appliances as they came off the assembly line to one of their warehouses for storage until the products were shipped to retail distributors across the nation. This arrangement worked well for both companies.

First marketed to the public in 1930, the refrigerator was redesigned in 1935 by the industrial designer Raymond Loewy and became a resounding commercial triumph. Sales of Seeger refrigerators that combined a Seeger cabinet with an electric compressor that was manufactured elsewhere increased from 65,000 for the 1934 model to 275,000 units in 1935.

Seeger and Sears were in the midst of a production and merchandizing transition in residential refrigeration—from the age of ice to an era of mechanical operations. Two years after forging the agreement with General Wood and Sears, Seeger shipped 308,551 refrigerators from its St. Paul plant. By 1934, 1,283,000 electric refrigerators of all makes had been manufactured and sold in the United States, while in 1941 the number had nearly tripled to 3,500,000. In 1940 the United States Census reported that among the nation's 34,672,413 occupied dwelling units, 14,949,967 or 43 percent of the accommodations had mechanical refrigerators. Following the introduction and the enormous success of the *Coldspot Super Six* in 1935, Seeger was forced to expand its facilities again to serve the manufacturing needs of a plant employing 2,000 workers and operating with three shifts twenty-four hours a day.

In 1920 the Seeger Refrigerator Company built 2,600 iceboxes. The entire plant that had produced this then amazing production total comprised just a part of the plant that was present by 1946—the space that occupied the Commercial Line and part of the Metal Shop. In those early years when iceboxes reigned, a small wood working department and the remainder of the Metal Shop occupied the rest of the ground floor with the cabinetmakers and assemblers on the second floor. The Finishing and Crating departments as well as the warehouse were on the third floor. A few hundred men comprised the entire work force building the cabinets on the floor and pushed the assembled iceboxes along on their own casters. Air supply for paint sprayers in these early days were so limited that often only one or two could operate their guns at the same time, the supply being furnished by a one-cylinder compressor of the portable type.

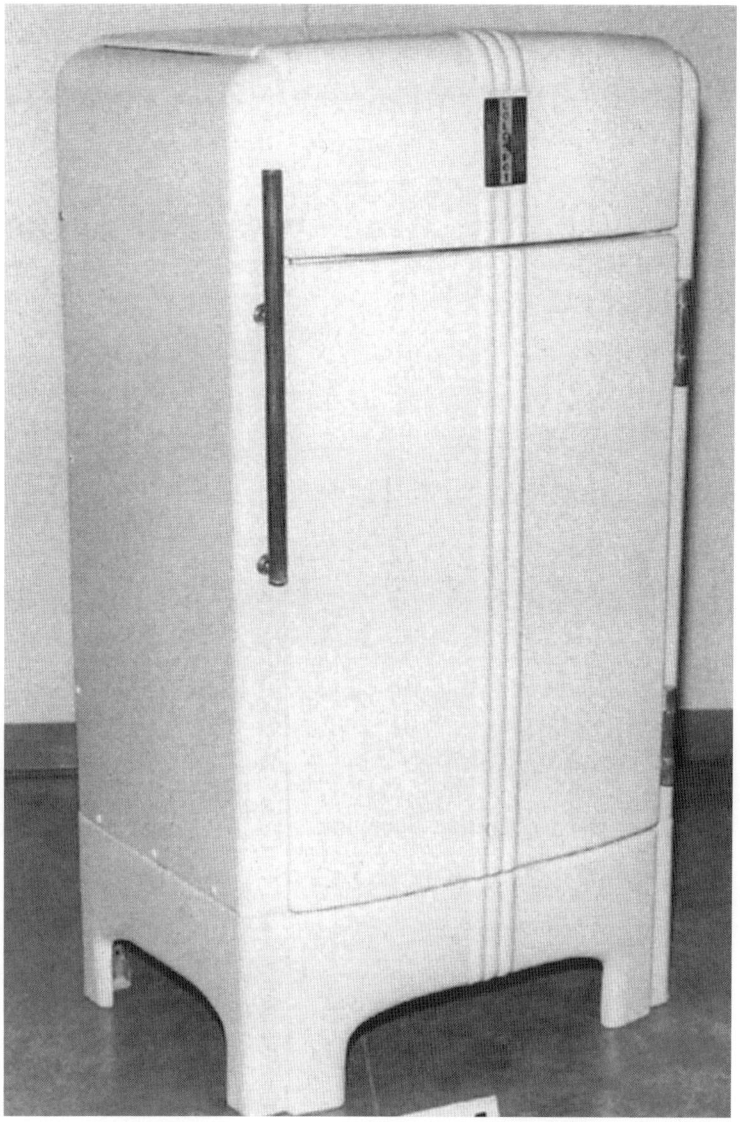

The celebrated Sears, Roebuck Coldspot Super Six, *which was designed by Raymond Loewy. Whirlpool Corporation photo.*

During the 1920s the company established sales offices in several strategic cities across the country to promote its products. From its headquarters in St. Paul, branch operations were set up in Boston, New York, and Atlanta on the East Coast; in Chicago in the heartland; and in Los Angeles and Seattle on the West Coast. Seeger refrigerators, ice machines, and other equipment soon appeared in private homes, restaurants, hotels, hospitals, clubs, schools and colleges at every point on the compass, from Yale University in New Haven, Connecticut, to the St. Paul Athletic Club, to the castle of newspaper mogul William Randolph Hearst at San Simeon in California.

The early 1920s witnessed the first of a series of expansion programs

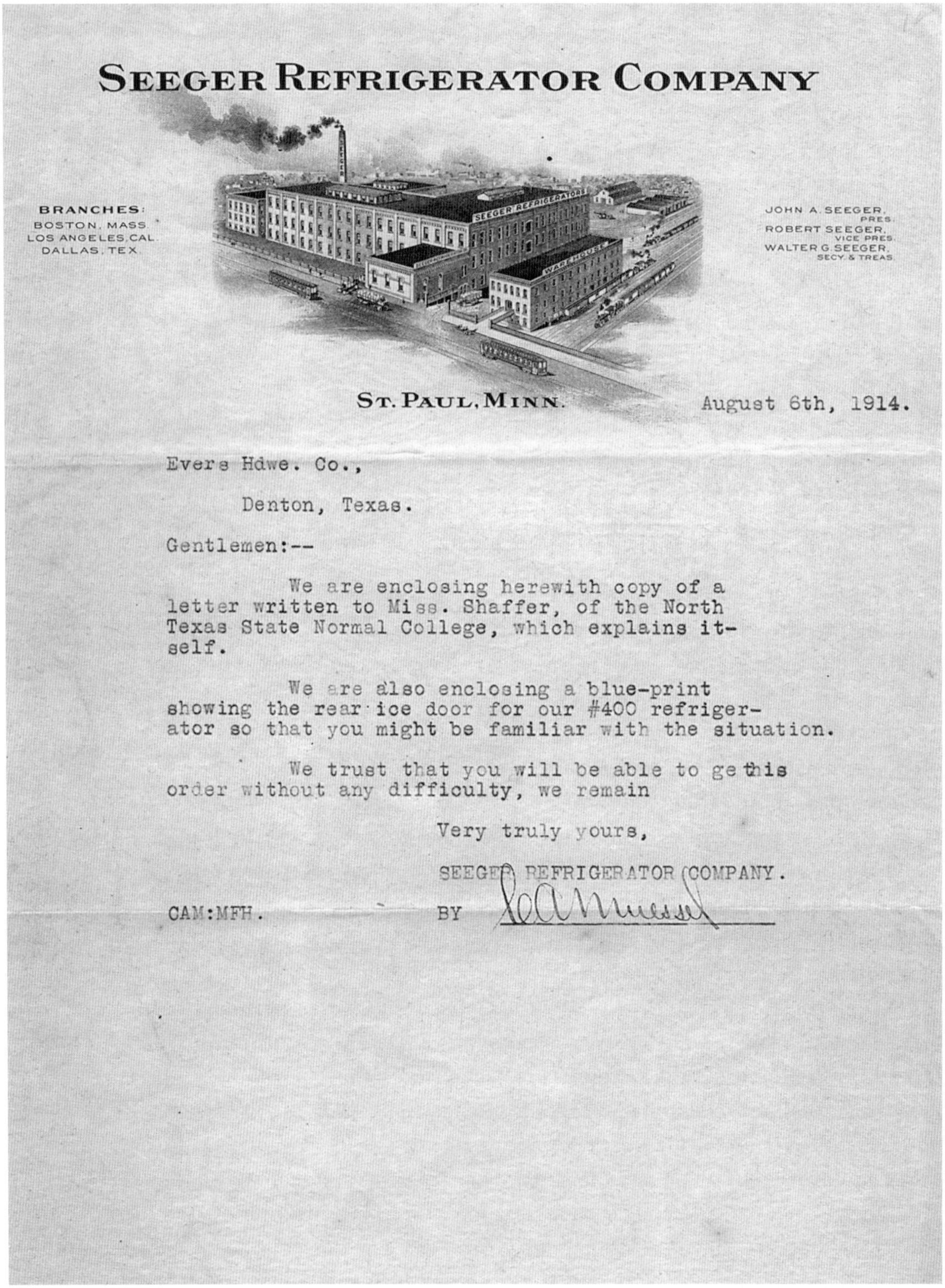

SEEGER REFRIGERATOR COMPANY

BRANCHES:
BOSTON, MASS.
LOS ANGELES, CAL.
DALLAS, TEX.

JOHN A. SEEGER, PRES.
ROBERT SEEGER, VICE PRES.
WALTER G. SEEGER, SECY. & TREAS.

ST. PAUL, MINN. August 6th, 1914.

Evers Hdwe. Co.,

Denton, Texas.

Gentlemen:--

We are enclosing herewith copy of a letter written to Miss. Shaffer, of the North Texas State Normal College, which explains it-self.

We are also enclosing a blue-print showing the rear ice door for our #400 refriger-ator so that you might be familiar with the situation.

We trust that you will be able to ge this order without any difficulty, we remain

Very truly yours,

SEEGER REFRIGERATOR COMPANY.

CAM:MFH. BY C A Muessel

A letter from Charles A. Muessel of the Seeger Company to a hardware store in Denton, Texas, regarding the purchase of a Seeger Refrigerator. The letter had a blueprint (reproduced here on the facing page) for the refrigerator in question to show how it would be installed in an institutional setting such as the North Texas State Normal College (today's North Texas State University in Denton). Letter and blueprint from the author's collection.

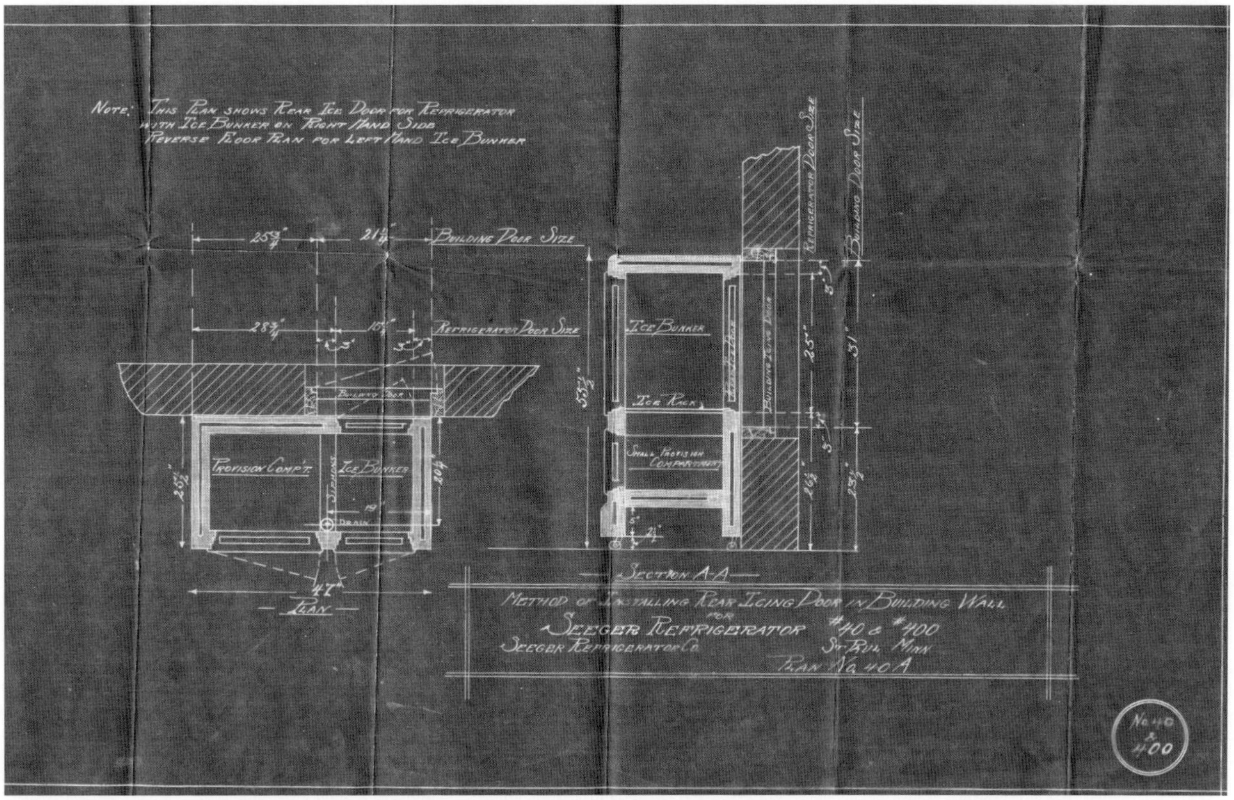

and manufacturing decisions that changed the market focus of the company from primarily a local industry with a regional clientele to a business that was national in its scope. The appearance of the electric refrigerator increased demands for porcelain in refrigerator cabinets. The foresight in discontinuing the manufacturing of ice refrigerators for household and commercial use, the development of a universal hanger for refrigerator units, and the concentration on the manufacture of cabinets for electric refrigerators proved to be vital factors in the company's growth.

Because Seeger had set up one of the first porcelain enameling departments in the nation and was strategically positioned for the new changes in the appliance marketplace, it was ready and capable of manufacturing boxes for electric refrigerators for thirty well-known home appliance manufacturers including General Electric, Kelvinator, Norge, Westinghouse, Copeland, Fairbanks Morse, and Amana.[11] The increased volume of orders that Seeger got from these manufacturers, who themselves were not as yet equipped to meet the increasing demand for their products in the now historical boom period of the 1920s, forced the Seeger Company to add considerably more manufacturing space and operating facilities to its then existing plant.

Seeger COMMERCIAL CABINETS

Hotels and Restaurants Equipped with Seeger Refrigerators

Hotel Des Artistes, New York City.
Victoria Hotel, Boston, Mass.
Hotel Wellington, Philadelphia, Pa.
Bristol Ferry Inn, Bristol Ferry, R. I.
The Castle Sq. Hotel, Boston, Mass.
The Dixville Notch Corp., Colebrook, N. H.
The Marshall House, York Harbor, Me.
St. Johnsbury Hotel, St. Johnsbury, Vt.
Colonial Hotel, Boston, Mass.
Iroquois Hotel, New York.
Parker House, New Bedford, Mass.
Hanover Inn, Hanover, N. H.
Bay of Naples Hotel, Naples, Me.
Greylock Hotel, Williamstown, Mass.
Kearsarge Hotel, N. Conway, N. H.
Saint Paul Hotel, St. Paul, Minn.
Suntaug Lake Inn, Lynnfield, Mass.
Narragansett Pier Hotel, Narragansett, R. I.
Lewis House, Beacon Falls, Conn.
Wentworth Hall Annex, Glen Jackson, N. H.
Ocean House, Swampscott, Mass.
The Balsams, Dixville Notch, N. H.
Chesire House, Keene, N. H.
Mt. Adams House, N. Woodstock, N. H.
Wentworth Hotel, Portsmouth, N. H.
Forest Hills Hotel, Franconia, N. H.
Riverside House, Green Harbor, Mass.
Alpine Hotel, N. Woodstock, N. H.
Narragansett Hotel, Rockland, Me.
Harbor View Hotel, Edgartown, Mass.
Chatham Bass Inn, Chatham, Mass.
Blue Pigeon, Bass Point, Mass.
Heidleberg Hotel, Boston, Mass.
Hotel Lucerne, Boston, Mass.
Kenmore Hotel, Boston, Mass.
Clarks Hotel, Boston, Mass.
Quincy House, Boston, Mass.
Hotel Garde, Hartford, Conn.
American House, Boston, Mass.
Hotel Warren, Worcester, Mass.
Falmouth Hotel, Portland, Me.
Samoset House, East Walpole, Mass.
New Kahler Hotel, Rochester, Minn.
Arlington Hotel, Boston, Mass.
Brun Hotel, St. Cloud, Minn.
Royal Lunch, Nos. 1 and 2, St. Paul, Minn.
Regal Lunch, Nos. 1, 2 and 3, St. Paul, Minn.
Trocke Cafeteria, St. Paul, Minn.
Viceroy Cafe, Minneapolis, Minn.
Hotel Lennox, Boston, Mass.
Tripp Memorial Building, Wisconsin University, Madison, Wis.
Lowry Hotel, St. Paul, Minn.
Saranac Lake Hotel, Saranac Lake, N. Y.
A. T. & S. F. Railway, Gallup, New Mexico.
Boston Y. W. C. A., Boston, Mass.
State Normal School, New Britain, Conn.
Y. W. C. A., Hartford, Conn.
Notre Dame Academy, Tynsboro, Mass.
The Hardware Club, New York City.
Savoy Plaza Apartment, New York City.
Rockefeller Institute, New York City.
W. T. Grant Company, Boston, Mass.
New York Central Station, New York, N. Y.
Edge Water Beach Hotel, Chicago, Ill.
St. Theresa College (Girls' Dormitory), Winona, Minn.
Phi Gama Delta, New York City.
Mutual Life Insurance, New York City.
Polly Royal Restaurant, South River, N. J.
Y. W. C. A., New York, N. Y.
Silver Bay Association, New York, N. Y.
West Side General Hospital, St. Paul, Minn.
Royal Netherlands West India Lines, New York City.
General Motors Building, New York, N. Y.
Congress Hotel, Chicago, Ill.
Legat Hotel, Butte, Mont.
Riggins Hotel, Waco, Texas.
Windsor Hotel, Wheeling, W. Va.
Arcade Cafeteria, Minneapolis, Minn.
Eat Shops, Nos. 1, 2, 3 and 4, St. Paul, Minn.
Radisson Hotel, Minneapolis, Minn.
C. & E. Mfg. Co., Restaurant, Chicago, Ill.
Hotel Florida, Havana, Cuba.
Congress Sq. Hotel, Portland, N. H.
Donaldson's Glass Block Restaurant, Minneapolis, Minn.
Emporium Tea Rooms, St. Paul, Minn.
Dayton Company, Minneapolis, Minn.
Gardner Hotel, Fargo, N. Dak.
Dispatch & Pioneer Press Lunch Co., St. Paul, Minn.
Glacier National Park Hotel, Glacier, Mont.
Hamline University Restaurant, Minneapolis, Minn.
Hastings Hotel, Minneapolis, Minn.
Fadden's Restaurant, St. Paul, Minn.
Roosevelt Hotel, New Orleans, La.
Kwong Tung Low Co., St. Paul, Minn.
Hotel Utah, Salt Lake City, Utah.

Hotels and Restaurants—(Cont.)

Miller Hotel Co., St. Paul, Minn.
J. A. Murphy Eating House, C. M. & St. P. Ry.
Lawton Mills Corp., Plainfield, Conn.
Mason Hotel, Corpus Christi, Texas.
Nueces Hotel, Corpus Christi, Texas.
Plaza Hotel, Minneapolis, Minn.
Jackson Hotel, Sioux City, Iowa.
Hotel Martin, Sioux City, Iowa.
U. S. Army, U. S. Navy, U. S. Steamship Saratoga.

Hospitals Equipped with Seeger Refrigerators

St. Mary's Hospital, Rochester, Minn.
Bridgeport Isolation Hospital, Bridgeport, Conn.
Nassau County Tuberculosis Hospital, Farmingdale, L. I., N. Y.
Emergency Hospital Buildings, New York Navy Yard, Brooklyn, N. Y.
Kingston Avenue Hospital, New York, N. Y.
St. Christopher's Hospital for Babies, Brooklyn, N. Y.
Rockland County Tuberculosis Hospital, Summit Park, N. Y.
New Rochelle Hospital, New Rochelle, N. Y.
Norwegian Deaconess Lutheran Hospital, Brooklyn, N. Y.
Mt. Kisco Hospital, Mt. Kisco, N. Y.
Wayside Home, Valley Stream, N. Y.
Framingham Hospital, Framingham, Mass.
Boston State Hospital, Boston, Mass.
U. S. Marine Hospital, Chelsea, Mass.
Hospital for Advanced Cases of Tuberculosis, Wallum Lake, R. I.
Barnstable County Infirmary, Pocassett, Mass.
House of Mercy Hospital, Pittsfield, Mass.
Long Island Hospital, Boston Harbor, Mass.
Connecticut State Hospital, Middletown, Conn.
Lowell Contagious Hospital, Lowell, Mass.
Lowell General Hospital, Lowell, Mass.
Salem Hospital, Salem, Mass.
Emergency Fleet Hospital, Boston, Mass.
Essex County Hospital, Middleton, Mass.
Crippled Children's Hospital, St. Paul, Minn.
Granite Falls Sanitarium, Granite Falls, Minn.
St. Joseph's Hospital, Mitchell, S. D.
Minnesota State Hospital, Hastings, Minn.
Minnesota State Asylum, St. Peter, Minn.
Minnesota State Hospital, Willmar, Minn.
St. Raphael's Hospital, St. Cloud, Minn.
St. Mary's Hospital, Minneapolis, Minn.
St. Mary's Hospital, Duluth, Minn.
Boston Lying-In Hospital, Boston, Mass.
Cambridge Municipal Hospital, Cambridge, Mass.
Glen Lake Sanitarium, Glen Lake, Minn.
Goodhue County Sanitarium, Red Wing, Minn.
Park Hospital, Livingston, Mont.
Cass County Hospital, Fargo, N. D.
Union Hospital, New Ulm, Minn.
City and County Hospital, St. Paul, Minn.
St. Luke's Hospital, St. Paul, Minn.
Minnesota State Hospital for Insane, Fergus Falls, Minn.
Minnesota State Hospital for Insane, Rochester, Minn.
Lutheran Deaconess Hospital, Minneapolis, Minn.
Elliott Hospital (State University), Minneapolis, Minn.
St. Paul Anti-Tuberculosis Hospital, St. Paul, Minn.
Swedish Hospital, Minneapolis, Minn.
Mounds Park Sanitarium, St. Paul, Minn.
Asbury Hospital, Minneapolis, Minn.
St. Luke's Hospital, Aberdeen, S. D.
Northern Pacific Hospital, Missoula, Mont.
Brooklyn Naval Hospital, Brooklyn, N. Y.
Chelsea Naval Hospital, Chelsea, Mass.
U. S. Naval Hospital, Charleston, N. C.
U. S. Medical Supply Depot, various hospitals throughout the United States.
Sisters of Notre Dame, Boston, Mass.
Deerwood Sanitarium, Deerwood, Minn.
Sunny Crest Sanitarium, Crookston, Minn.
Fair Oaks Sanitarium, Wadena, Minn.
St. Joseph's Hospital, Minot, N. D.
Oakland Park Sanitarium, Thief River Falls, Minn.
Women's Reformatory, Shakopee, Minn.
Worrell Hospital, Rochester, Minn.
State School for Feeble Minded, Faribault, Minn.
Miller Hospital, St. Paul, Minn.
St. Joseph's Hospital, St. Paul, Minn.
The New St. Joseph's Hospital, St. Paul, Minn.
Geo. Geissinger Memorial Hospital, Danville, Pa.

Hospitals Equipped with Seeger Refrigerators—(Cont.)

Mary Immaculate Hospital, New York City.
Lutheran Hospital, Brooklyn, N. Y.
Walter Reed General Hospital, Washington, D. C.
Lincoln Hospital, New York City.
Columbia Presbyterian Hospital, New York City.

Clubs

Columbia University Club, New York City.
Central Club for Nurses, New York City.
Mt. Kisco Gulf Club, Mt. Kisco, N. Y.
Shuttle Meadow Club, New Britain, Conn.
Boston Athletic Club, Boston, Mass.
Long Sought For Lodge, Westford, Mass.
Wianna Club, Osterville, Mass.
Iroquois Club, Cambridge, Mass.
Woman's City Club, Boston, Mass.
Kernwood County Club, Salem, Mass.
Myopia Hunt Club, Hamilton, Mass.
Essex Club, Essex, Mass.
Phi-Beta Epsilon Corp., Cambridge, Mass.
Boston Yacht Club, Nantasket Beach, Mass
Y. M. C. A., Hartford, Conn.
Y. W. C. A., Boston, Mass.
Fraternal Order of Eagles, Rockland, Me.
Yale University Club, New Haven, Conn.
Cumberland Club, Portland, Me.
Town and Country Club, St. Paul, Minn.
Minneapolis Athletic Club, Minneapolis, Minn.
St. Paul Athletic Club, St. Paul, Minn.
White Bear Yacht Club, White Bear, Minn.
Automobile Club, St. Paul, Minn.
Capital City Club, Atlanta, Ga.
Elk's Club, St. Paul, Minn.
Elk's Club, Minneapolis, Minn.
Elk's Club, Kenosha, Wis.
Minneapolis Commercial Club, Minneapolis Minn.
Elk's Club, Havre, Mont.
Hamilton Club, Chicago, Ill.
Belmont Springs Country Club.
Greenbrook Country Club, Jersey City, N. J.
Ridgewood Country Club, Ridgewood, N. J.
Riverview Corp., Yonkers, N. Y.
Annandale Club, Los Angeles, Cal.
Minnesota Club, St. Paul, Minn.
Delta Upsilon, Minneapolis, Minn.
Watatic Club, Ashburnham, Mass.
Babson Club, Wellesley Hills, Mass.
Weston Golf Club, Weston, Mass.
Wisconsin Club, Milwaukee, Wis.
University Club, Pasadena, Cal.
The Dayton Club, Dayton, Ohio.

Schools and Colleges

Sisters of Notre Dame, Boston, Mass.
State Normal College, Bemidji, Minn.
High School of Commerce, Springfield, Mass.
Milton Academy, Milton, Mass.
Girls' Trade School, Cambridge, Mass.
Commercial College for Women, Hartford, Mass.
Dedham High School, Dedham, Mass.
State Normal School, Bridgewater, Mass.
Commerce School for Girls, Springfield, Mass.
Massachusetts Agricultural College, Amherst, Mass.
Walnut Hill School, Natick, Mass.
Phillips Exeter School, Exeter, N. H.
English High School, Waterbury, Conn.
East Somerville Junior High School, Somerville, Mass.
Winter Hill Junior High School, Somerville Mass.
Vassar College, Poughkeepsie, N. Y.
Coast Guard Academy, New London, Conn.
South High School, Minneapolis, Minn.
West High School, Minneapolis, Minn.
Mechanic Arts High School, St. Paul, Minn.
Central High School, St. Paul, Minn.
Johnson High School, St. Paul, Minn.
St. Paul Academy, St. Paul, Minn.
State Normal School, St. Cloud, Minn.
State Training School, Red Wing, Minn.
Dunwoodie Institute, Minneapolis, Minn.
Wellesley College, Wellesley, Mass.
Carleton College, Northfield, Minn.
Hamline University, St. Paul, Minn.
Moorhead Normal, Moorhead, Minn.
Anoka College, Anoka, Minn.
School for Deaf, Faribault, Minn.
Winona Normal School, Winona, Minn.
Alice Shevlin Hall, Minnesota University, Minneapolis, Minn.
Alhambra High School, Alhambra, Cal.
Monrovia High School, Monrovia, Cal.

[*Twenty-four*]

This is a Seeger Company list of its commercial clients reproduced from one of their catalogs in the 1920s or '30s. From a Seeger catalog in the author's collection.

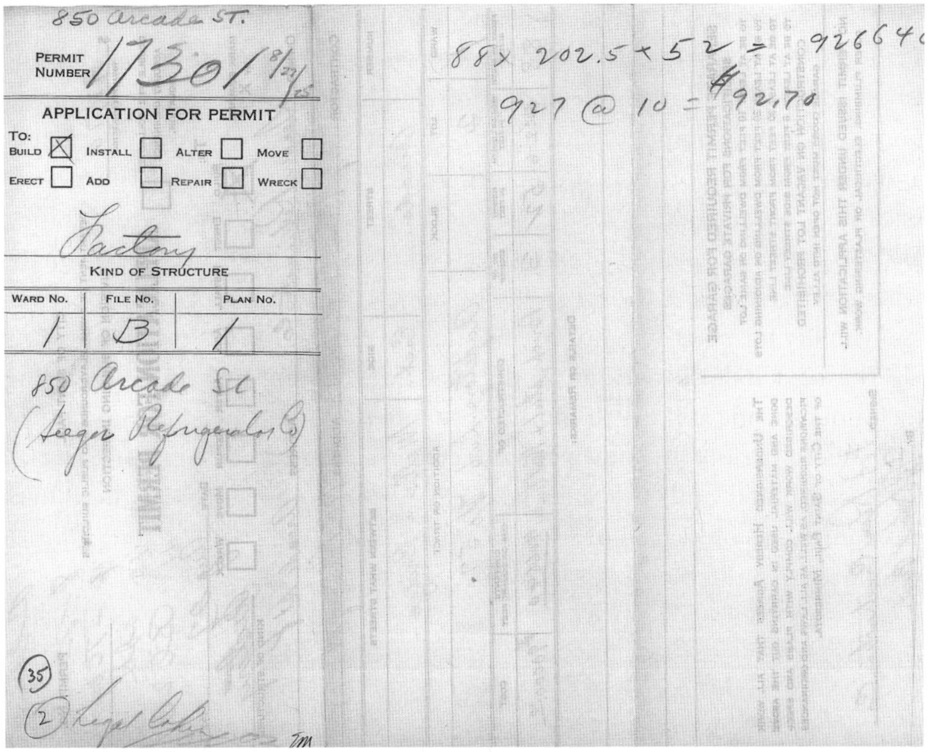

850 Arcade St.

PERMIT NUMBER 17301 8/27/25

88 x 202.5 x 52 = 92664

927 @ 10 = $92.70

APPLICATION FOR PERMIT

TO: BUILD [X] INSTALL [ ] ALTER [ ] MOVE [ ]
ERECT [ ] ADD [ ] REPAIR [ ] WRECK [ ]

Factory

KIND OF STRUCTURE

WARD NO.	FILE NO.	PLAN NO.
1	B	1

850 Arcade St
(Seeger Refrigerator Co)

The outside of the Seeger Company's August 1925 application for a building permit for a major addition to its plant at 850 Arcade Street.

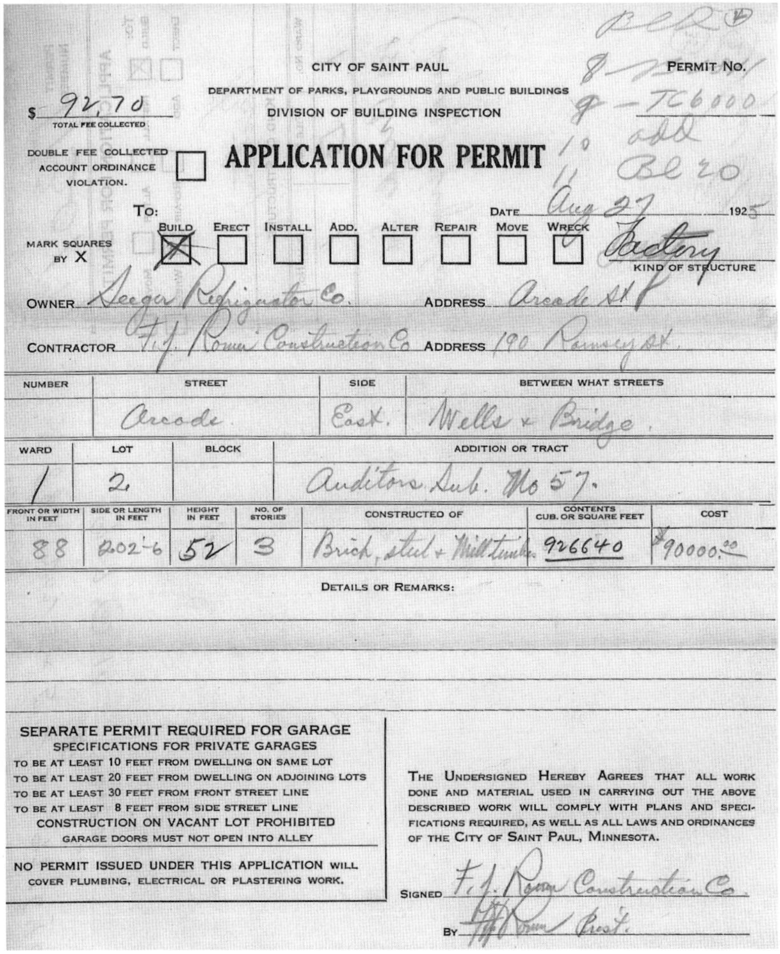

CITY OF SAINT PAUL
DEPARTMENT OF PARKS, PLAYGROUNDS AND PUBLIC BUILDINGS
DIVISION OF BUILDING INSPECTION

PERMIT No.

$ 92.70 TOTAL FEE COLLECTED

DOUBLE FEE COLLECTED ACCOUNT ORDINANCE VIOLATION. [ ]

APPLICATION FOR PERMIT

TO: DATE Aug 27 1925

MARK SQUARES BY X: BUILD [X] ERECT [ ] INSTALL [ ] ADD. [ ] ALTER [ ] REPAIR [ ] MOVE [ ] WRECK [ ]

Factory KIND OF STRUCTURE

OWNER Seeger Refrigerator Co. ADDRESS Arcade St

CONTRACTOR F.J. Romer Construction Co ADDRESS 190 Ramsey St.

NUMBER	STREET	SIDE	BETWEEN WHAT STREETS
	Arcade	East.	Wells + Bridge.

WARD	LOT	BLOCK	ADDITION OR TRACT
1	2		Auditors Sub. No 57.

FRONT OR WIDTH IN FEET	SIDE OR LENGTH IN FEET	HEIGHT IN FEET	NO. OF STORIES	CONSTRUCTED OF	CONTENTS CUB. OR SQUARE FEET	COST
88	202-6	52	3	Brick, steel + Mill timb.	926640	$90000.00

DETAILS OR REMARKS:

SEPARATE PERMIT REQUIRED FOR GARAGE
SPECIFICATIONS FOR PRIVATE GARAGES
TO BE AT LEAST 10 FEET FROM DWELLING ON SAME LOT
TO BE AT LEAST 20 FEET FROM DWELLING ON ADJOINING LOTS
TO BE AT LEAST 30 FEET FROM FRONT STREET LINE
TO BE AT LEAST 8 FEET FROM SIDE STREET LINE
CONSTRUCTION ON VACANT LOT PROHIBITED
GARAGE DOORS MUST NOT OPEN INTO ALLEY

NO PERMIT ISSUED UNDER THIS APPLICATION WILL COVER PLUMBING, ELECTRICAL OR PLASTERING WORK.

THE UNDERSIGNED HEREBY AGREES THAT ALL WORK DONE AND MATERIAL USED IN CARRYING OUT THE ABOVE DESCRIBED WORK WILL COMPLY WITH PLANS AND SPECIFICATIONS REQUIRED, AS WELL AS ALL LAWS AND ORDINANCES OF THE CITY OF SAINT PAUL, MINNESOTA.

SIGNED F.J. Romer Construction Co.
BY F.J. Romer Pres.

The reverse side of the Seeger building permit application indicates that the addition was to be 88 feet wide, 202 feet deep, and 52 feet high. It would be three stories high, constructed of brick, steel, and mill timber materials, with an estimated cost of $90,000. St. Paul Building Permits Collection, Ramsey County Historical Society, St. Paul, Minn.

The second major physical changes to the Seeger plant took place in 1924 and 1925. A large steel warehouse was erected for the storage of finished goods, a new assembly and wood mill building was added, a modern porcelain shop, equipped with all the newest equipment was constructed, and a small neighboring manufacturing plant was purchased.[12] The factory became more mechanized and electrically driven conveyers were installed and production reached about sixty cabinets a day, an accomplishment which amazed John A. Seeger, who by now was placing more and more responsibility for plant operations and management upon his two sons.

The next half decade found the plant developing new manufacturing techniques while still continuing as a supplier of domestic and commercial cabinets to nationally known manufacturers. These client-vendor associations continued until the sales volumes of these appliances at the different companies increased to such an extent that these clients found it more profitable to build their own factories.

3

Refrigeration Progress: An Era of Residential and Commercial Change

TODAY, IN THE KITCHEN of nearly every home in America—more than 99.5 percent of households—there is a refrigerator. It is an appliance that is as commonly accepted as the light bulb and the electric toaster, but it is also one of those miracles of modern living that totally changes the condition of everyday life. In the course of domestic history, the refrigerator is a recent development in a long chain of technological advancements particularly from the late nineteenth century to the present.[1]

John A. Seeger, founder and chairman of the Seeger Refrigerator Company, at the entrance to the Arcade Street plant about 1935. Photo from the author's collection.

The innovations in refrigeration technology and engineering in the twentieth century pioneered elsewhere were paralleled in St. Paul, Minnesota, by the founding of the Seeger Refrigerator Company in 1902 and its subsequent expansion based on timely exploitation of technological innovations both in the engineering of cooling systems and in the materials the company used in the construction of its appliances. Over five decades John A. Seeger and his sons, Walter and Gustave, with a team of colleagues, transformed a small, fledgling company that was selling to a local market into a large manufacturing firm capable of reaching consumers nationwide. They did this through the introduction of improved manufacturing procedures and by making timely appliance design changes that made the Seeger Company a key producer in the national appliance market.

People store foods today in their home refrigerators and freezers. Supermarkets use refrigerated display cases; they also have freezing rooms and cold-storage warehouses. Fresh foods are carried long distances in refrigerated trucks, railway cars, and refrigerated compartments of ships. On camping trips people keep food fresh for days with portable refrigerators.

Air conditioning depends on refrigeration to cool homes, offices, theaters, stores, and automobiles and makes it possible to store serums, vaccines, and blood plasma, among other lifesaving medical supplies. Drug companies use refrigeration to make penicillin and other drugs. Furs are stored in refrigerated vaults to protect them. Florists refrigerate cut flowers; drinking fountains supply cold water; and ice machines provide blocks, cubes, flakes, and chips of ice. Skating rinks and ice plants use refrigerating machines to make ice. Industrial companies around the world use refrigeration to process rubber, lubricants, steel, frozen fruit juices, candy, photographic films, ice cream, chemicals, and so forth.

The household refrigerating machine had been under active development since the middle of the nineteenth century. By 1920 various refrigerating machines were being manufactured in sufficient quantities to prove that the appliance had the potential to be a commercial success. Considerable progress had been made in producing better machines, ones that were more reliable in operation, of more simplified construction, and lower in initial cost.

Mechanical household refrigeration needs had an important influence on refrigerator cabinet construction during this time. These refrigerators had to have better construction and insulation to operate satisfactorily in homes. This necessitated the design and construction of an appliance that was capable of consistently providing the lower food compartment temperatures produced by a mechanical unit. Thus food compartment temperatures of 40 degrees Fahrenheit were unquestionably more desirable than the higher temperatures that had previously been considered satisfactory with older cooling technologies.

The sales of household refrigerators and freezers beginning in 1920 that are given in Figure 1 demonstrate just how rapidly the market for these appliances expanded once the engineering and construction of them improved. Figure 1 also documents how the average retail price of household refrigerators and freezers came down as sales went up between 1920 and 1950 and how refrigerator manufacturers had to grow to keep up with consumer demand.

The various improvements in household refrigerators together with the decrease in electric rates that was taking place across the United States at this time resulted in a cost of operation usually favorable to the machine when compared to the cost of ice. By the 1930s accessories such as ventilated vegetable pans, temperature regulators, electric lights, sliding shelves, water coolers, and defrosting devices were useful features of convenience that leading manufacturers included in many of their models.

FIGURE 1

Household Refrigerators and Freezers

Summary World Household Refrigerator Sales of U.S. Manufacturers

YEAR	UNIT SALES	AVERAGE RETAIL PRICE	RETAIL VALUE	WIRED HOMES	REFRIGERATORS IN USE	SATURATION
Up to						
1920	10,000	$600	$ 6,000,000			
1921	5,000	550	2,750,000			
1922	12,000	525	6,300,000			
1923	18,000	475	8,550,000			
1924	30,000	450	13,500,000			
1925	75,000	425	31,875,000			
1926	210,000	390	81,900,000			
1927	390,000	350	136,500,000			
1928	560,000	334	187,040,000			
1929	840,000	292	245,280,000			
1930	850,000	275	233,750,000			
1931	965,000	258	248,970,000			
1932	840,000	195	163,800,000			
1933	1,080,000	170	183,600,000			
1934	1,390,000	172	239,080,000			
1935	1,688,000	166	280,208,000			
1936	2,180,000	164	357,520,000			
1937	2,500,000	171	427,500,000			
1938	1,410,000	172	242,520,000			
1939	2,085,000	169	352,365,000	24,450,000	13,740,000	57%
1940	2,720,000	154	418,880,000	26,000,000	15,090,000	58%
1941	3,700,000	160	592,000,000	26,800,000	19,090,000	71%
1942	520,000	165	86,000,000	27,600,000	19,410,000	71%
1943	**	*	**	27,600,000	19,350,000	70%
1944	**	*	**	27,600,000	19,250,000	69%
1945	263,000	180	47,400,000	29,215,000	19,220,000	66%
1946	2,100,000	207	434,700,000	31,015,000	20,900,000	67%
1947	3,400,000	240	816,000,000	33,500,000	23,600,000	70%
1948	4,800,000	250	1,202,500,000	35,205,000	27,200,000	76%
1949	4,450,000	251	1,134,750,000	37,299,100	29,500,000	79%
1950	6,200,000	258	1,602,266,000	39,044,000	33,750,000	86%
1951	4,075,000	274	1,120,625,000	40,967,180	35,500,000	87%
1952	3,570,000	285	1,017,450,000	42,306,598	37,750,000	89%
1953	3,775,000	295	1,113,625,000	43,590,000	39,397,000	90%

Refrigeration Diary and Reference Book, 1955 (Tecumseh, Michigan, 1954): 59.

FIGURE 2

Freezer Sales by Years

YEAR	UNITS	AVERAGE RETAIL PRICE	$ VOLUME	SATURATION
1946	210,200	$380	80,000,000	.6
1947	607,000	335	200,310,000	2.3
1948	690,000	320	220,800,000	4.1
1949	485,000	335	162,475,000	5.2
1950	809,000	325	289,250,000	7.2
1951	1,050,000	360	378,000,000	9.3
1952	1,140,000	370	421,800,000	11.5
1953	1,200,000	400	480,000,000	13.7

Refrigeration Diary and Reference Book, 1955 (Tecumseh, Michigan, 1954): 59.

About two-thirds of the homes in the United States were electrically wired by the mid-1930s. Consequently there were about 21,000,000 domestic electric customers. About 17 percent of these wired homes had electric refrigerating machines in 1935, indicating that there was a large potential market for refrigerators. About 12,000,000 refrigerators that used ice as their refrigerant were in use in the United States at the same time.

Figure 1 provides the annual production of household refrigerators during the 1920s and early 1930s. Figure 2 gives a snapshot of freezers sales in the first eight years after World War II, a period of high consumer demand for all types of household goods:

As there was little electricity available for private consumption in the 1890s and early years of the twentieth century, there were no electrically powered refrigerators. In most homes an icebox was used to store perishables in the summer. On farms that had an icehouse, farmers typically had an icebox. In the first decades of the twentieth century, the majority of Americans did not live in cities, they lived on farms; and the majority of those did not have electricity until World War II.

Iceboxes date back to the 1830s. Most iceboxes were insulated with mineral wool, charcoal, cork, sawdust, flax straw fiber, or seaweed. The inside of the icebox was usually lined with galvanized metal, tin, zinc, slate, porcelain, or wood. The outside was finished in oak, pine, ash, or metal.[2] The average home icebox would hold a 25- or 50-pound block of ice. The price of a 25-pound block was about 15 cents in the early 1900s. The ice block would last one or two days during the hot summer months. A small drain built into the icebox would direct the melted ice water into a pan underneath the icebox. This pan had to be emptied frequently to avoid getting water all over the kitchen floor.

Iceboxes work because warm air rises. A cake of ice in the upper part of an icebox absorbs heat from the warm air. This cools the warm air and increases its density. The heavier air flows downward to the food compartments. The air becomes warmer and lighter as it absorbs heat from the food. The warmer, lighter air rises and again loses heat to the ice. Around 1913 electric refrigerators started to be mass-produced. Aggressive sales tactics by the refrigerator companies in the 1920s and 1930s contributed to the gradual demise of the icebox.

Seeger Siphon Refrigerators

Dry Air and its Circulation

THE illustration will help you to fully comprehend the scientific construction and operation of the *Seeger Siphon System* original patent and gives you some idea of the part it plays in refrigeration. The siphons when assembled are in one piece and may be placed at the side, end or rear of the refrigerrator, but they always form the partition between the ice and provision chambers, extending down to the deflector. The ice is placed on an ice grate, leaving a space underneath for the cold air to pass through and force its way into the provision chamber, rising through the openings provided in the wire shelves.

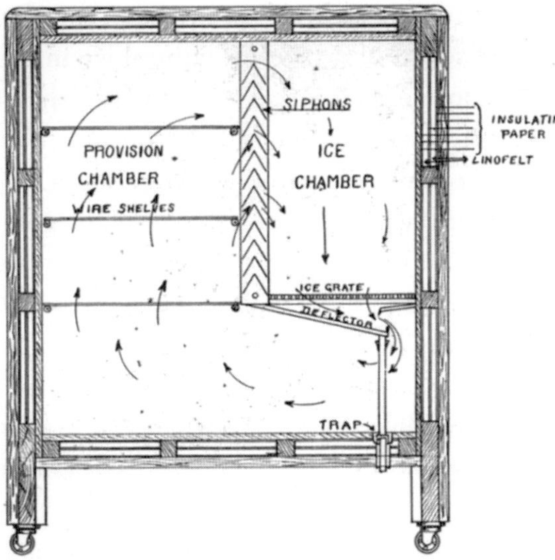

The patent siphon then draws the warmer air from the provision chamber into the ice chamber, carrying with it all odors and impurities which are condensed and carried off in the melted ice through the drip pipe.

As you will readily see, by this system all odors and gases are carried off by the strong and continuous current of dry, fresh air passing through the provision chamber, *permitting all kinds of fruits, vegetables, onions and melons to be stored in the provision chamber without affecting the cream, milk or butter.*

Page Four

A diagram from an early Seeger catalog demonstrating how the Seeger Original Sipher System worked in an icebox. From a Seeger catalog in the author's collection.

First, a bit of necessary etymology. The verb "to refrigerate" simply means to cause to become cold," with recorded uses as far back as the sixteenth century. The noun "refrigerator," meaning that which refrigerates or cools (in early use) or, in later use, an appliance, vessel, or chamber for maintaining a low degree of temperature, was popularized in its modern usage by a Maryland engineer named Thomas Moore. There are two classes of refrigerators: those that do not use a power source at the point of refrigeration, and those that do. Examples of the first category would include such processes as a hole in the ground or root cellar; a stand-alone icebox that operated with either natural or manufactured ice; and water evaporation devices. The examples of refrigerators that use a power source include the electrical and the gas-driven appliance.

The term "refrigerator" as we know it today came into common usage after Thomas Moore built the first appliance for refrigeration in 1800. Moore's device would nowadays be called an "icebox"—a cedar tub insulated with rabbit fur, filled with ice that surrounded a container for transporting butter from rural Maryland to Washington, D.C. By 1890 there were many refrigerators of various types manufactured in every community of any size that used naturally occurring ice. But warm winters in 1889 and 1890 created severe shortages of natural ice in the United States. This stimulated the commercial development of mechanical refrigeration for the freezing and storage of fish and in the brewing, dairy, and meat packing industries. This we shall call "artificial refrigeration."

Pioneers in artificial refrigeration during the eighteenth and nineteenth centuries included Dr. William Cullen, a Scotsman whose studies in the early 1700s dealt with the evaporation of liquids in a vacuum. Cullen demonstrated the first known artificial refrigeration in 1748 at the University of Glasgow. He did not use his discovery, however, for any practical purpose. In 1805, an American inventor, Oliver Evans, designed the first refrigeration machine and an American physician, John Goorie of Apalachicola, Florida, built a refrigerator based on Evans's design in 1834 to make ice to cool the air for his yellow fever patients. Other inventors who made extraordinary contributions to the modern refrigeration process were Michael Farady (1791–1867), a Londoner who in the early 1800s liquefied ammonia to cause cooling, and German engineer Carl von Linden, who in 1876 patented not a refrigerator but the process of liquefying gas that is part of basic refrigeration technology.[3]

Today's compression refrigeration system operates on a concept adapted from Farady's experiments. It involves compressing gas into a liquid which when allowed to expand back into a gas, will absorb heat. This is still valid as a simplified description of what happens in a home refrigerator, freezer, air conditioner, or dehumidifier.[4]

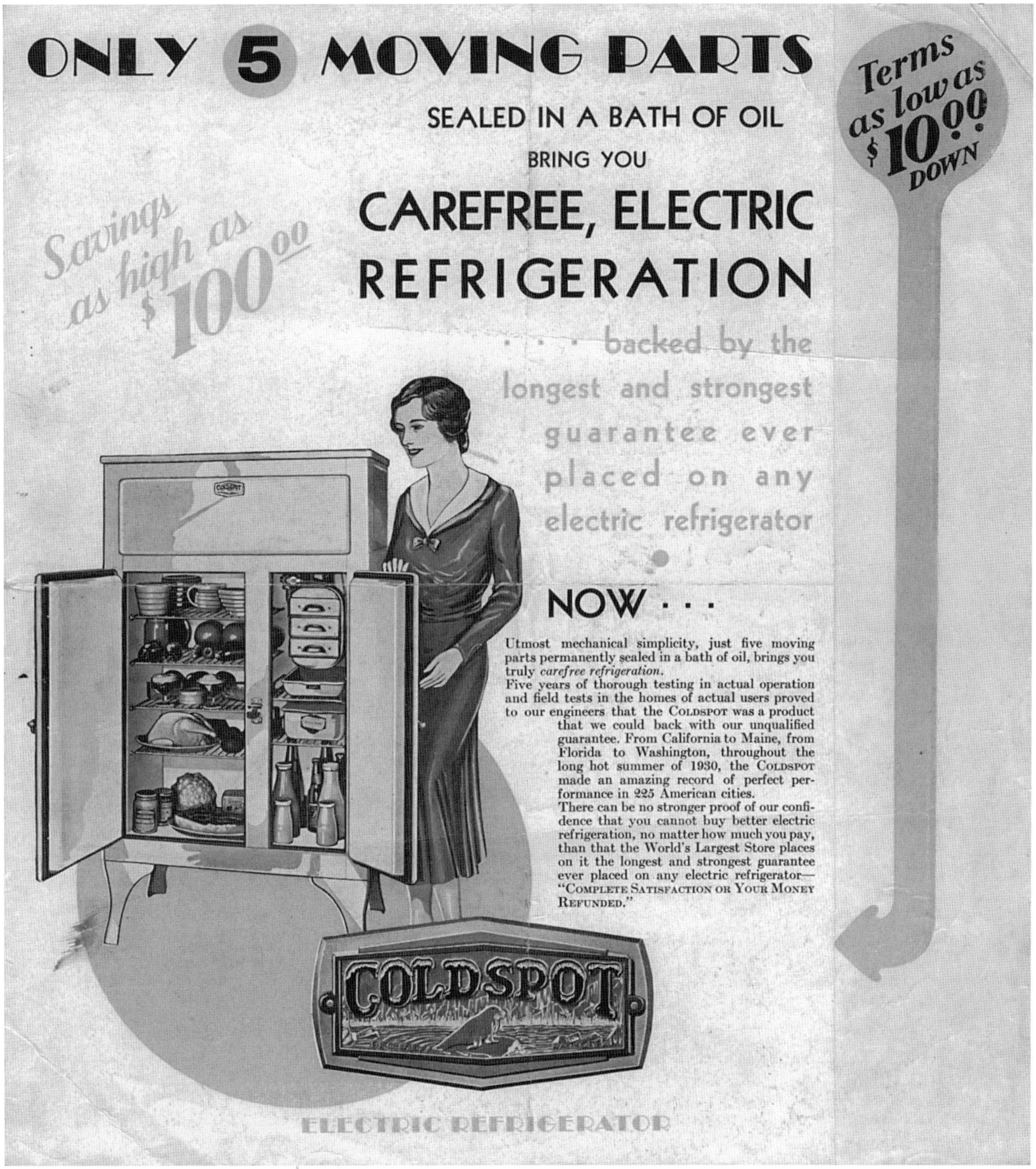

A Sears, Roebuck advertisement from 1930 for the first in the Sears line of Coldspot *refrigerators. The Seeger Company made the cabinet for this refrigerator. From a Sears, Roebuck advertisement in the author's collection.*

A Seeger advertisement from the 1920s in which either gas or electricity is used to power the refrigerator. It also promotes the "Chiltray" for fruit or vegetable storage. From a Seeger catalog in the author's collection.

Snow and ice, cool streams, springs, caves, and cellars were long ago used to refrigerate food. Meat and fish were preserved in warm weather by salting or smoking. The Chinese cut and stored ice as early as 1,000 B.C. and around 500 B.C. the Egyptians and Indians made ice on cold nights by setting water out in earthenware pots and keeping the pots wet. At some uncertain period in history, perhaps in fourteenth century China or seventeenth century Italy, it was discovered that the evaporation of brine (salt water) absorbed heat and therefore a container placed in brine would stay "cold." In eighteenth century England, servants collected ice in the winter and put it into icehouses, where the sheets of ice were packed in salt, wrapped in strips of flannel, and stored underground to keep them frozen until summer. At the beginning of the nineteenth century, iceboxes were used in England.

Natural ice was harvested, distributed and used in both commercial and home applications in the mid-1800s.[5] People cool with ice if they lack convenient power to produce other methods of refrigeration. Natural ice, cut from lakes and ponds in winter, has long provided refrigeration during warm seasons.[6]

Ice refrigerates because it absorbs heat when it melts. For example, this absorption happens when we cool a warm drink by putting ice cubes in the glass. Ice makes a useful refrigerant because it has a constant melting temperature of 32 degrees Fahrenheit (O degrees Centigrade). It absorbs large quantities of heat as it melts. The ice that has not melted, however, always maintains the same temperature. Ice is used to cool foods in iceboxes or to freeze liquids by *endothermic reactions*. These chemical reactions enable ice to produce temperatures below freezing.

In the United States, two of the first electric home refrigerators both appeared in Fort Wayne, Indiana, where, in 1911, the General Electric Company unveiled a unit invented by a French monk. In 1915 the first "Guardian" model—a predecessor of the Frigidaire—was assembled in a washhouse in a Fort Wayne backyard. The next year the Kelvinator and Servel models were among more than two dozen home refrigerators introduced by different manufacturers to the United States market and by 1920 the number of refrigerator types had increased to more than 200. The refrigerator compressors in these early appliances were generally driven by belts attached to motors that were located in the home's basement or in an adjoining room.

In 1918 the makers of the Kelvinator model introduced the first refrigerator with any type of automatic control. One manufacturer's 1922 model had a wooden cabinet, a water-cooled compressor, two ice cube trays, and nine cubic feet of storage space. The other type of innovation in refrigeration in these years concerned materials technology. Steel and porcelain cabinets began appearing in the mid-1920s. Refinements and improvements in refrigerators occurred regularly.

In the 1920s and '30s consumers were introduced to freezers when the first electric refrigerators with ice cube compartments came on the market and Freon 12 (the trademark for a colorless, odorless, nonflammable, noncorrosive, and nontoxic gas or liquid that is made up of fluorine and carbon) was used to replace sulfur dioxide as the most commonly used refrigerant. Prior to the introduction of Freon as a refrigerant in the late 1920s, the older refrigerants that were then in use (ammonia, methyl chloride, and sulfur dioxide, all of which are toxic gases) had caused a number of fatal accidents when they leaked out of their cooling systems. Consequently some householders had put their refrigerators in their backyards, an inconvenience when preparing food.

During the 1940s frozen food storage became widely used by consumers. Refrigeration technology increased in tempo in the 1950s and '60s when innovations like automatic defrost and automatic icemakers first appeared. By 1950 the Seeger Company, with plants in St. Paul, and Evansville, Indiana, manufactured a line of domestic and commercial refrigeration units, freezers, and dehumidifiers at the rate of 815,000 units per year. At this time Seeger employed more than 6,100 people.

4

1928: An Alliance with Sears Begins

In the annals of the Seeger Refrigerator Company, 1928 was a landmark year no less significant for the company's future accomplishments than 1902 was for its founding. It was the era of the "booming twenties," a decade marked by a vibrant national economy in nearly all sectors except agriculture coupled with dynamic technological, scientific, and industrial advancements. The election of Herbert Hoover as U.S. president in November of 1928 seemed to be a sign that the country was destined for a long period of sustained prosperity.

After more than twenty-five years of operation, Seeger was a mature, well-organized refrigerator manufacturing corporation, comprising many functional departments. Since its establishment, the Seeger Company had earned a distinguished national reputation for the quality of its wooden residential and commercial iceboxes. Beginning in 1916 it manufactured the steel cabinets for refrigerators that carried the label of the General Electric Corporation. In the 1920s other appliance firms, such as Kelvinator, Norge, Westinghouse, Copeland, Servel, and Fairbanks-Morse, would follow GE's example, installing their own electric or gas compressors in Seeger cabinets and applying their own brand name to the refrigerator.[1]

An early 1930s Seeger refrigerator. Photo courtesy of the Whirlpool Corporation.

For about four or five years from 1926 until 1930, Sears, Roebuck and Company merchandized a *Coldspot* electric refrigerator manufactured by the Chattanooga-based Tennessee Furniture Company.[2] By 1928, however, the Seeger Company began developing and manufacturing a similar appliance for Sears. It remains unclear how or who made the first contact between Seeger and Sears, Roebuck and Company to discuss the refrigerator to be sold by the merchandiser. Perhaps it was Walter G. Seeger attempting to establish a link with the Chicago-based company or possibly it was General Robert E. Wood, the chairman of

Sears, seeking to explore a business relationship. In any event 1928 signaled the beginning of a long-term association between the companies for the manufacture and sales of Seeger-made electric refrigerators.

Why did Sears turn to St. Paul's Seeger Company for the manufacture of its private brand appliance in 1928? The answer to the query seems embedded in the company's more than twenty-five-year history. Seeger had a strong and positive reputation for manufacturing quality cabinets for General Electric and other firms for more than ten years and it had been producing its own brand electric refrigerator for at least two years. Sears had had for sometime been a customer of the Sunbeam Electric Corporation in Evansville, Indiana, its supplier of electric generators for home and farm use. The three companies agreed in 1928 that Seeger would join with Sunbeam to develop and manufacture the first *Coldspot* refrigerators for Sears—Seeger supplying the cabinet and Sunbeam the electric motor and compressor to operate the appliance.

Nowadays, complex and signed contracts are the normal practice among corporations, government agencies, and individuals; the Seeger experience and association with Sears was different. The symbols of the working ties

General Robert E. Wood

Born June 13, 1879, in Kansas City, Missouri, Wood graduated from the U.S. Military Academy at West Point in 1900. In 1905, after serving two years in the Philippines, he was ordered to the Panama Canal Zone, where he worked for the next ten years. In May 1915, Wood retired from the army to enter business. When the United States entered World War I in 1917, Wood immediately offered his services to the War Department and served as a colonel in the 42nd (Rainbow) Division. He later assumed duties as the acting quartermaster general, headquartered in Washington, D.C., with the rank of brigadier general. Wood returned to civilian life in 1919 and joined Montgomery Ward, the mail-order giant, as its general merchandise manager before being named vice president. General Wood left the company in 1924 to become vice president of Sears and to lead its factory operations. He became president of Sears in January 1928 and chairman in 1939. Under Wood's leadership, Sears went from being the largest mail-order business primarily serving the rural population of the United States to the world's largest merchandiser. He retired as chairman of Sears in 1954 but continued as a director of the company until 1968. He died November 6, 1969.

General Robert E. Wood in 1941. Photo courtesy of Bettmann / CORBIS. © Bettmann / CORBIS.

between John A. and Walter G. Seeger with General Robert E. Wood of Sears, Roebuck and Company, were merely their word and a handshake. That simple, trusting, and honorable manner of doing business has largely passed from the scene with the icebox. The association with Sears was remarkable for its ethical base and its duration, and remains so today. Possibly it is the longest standing such association between a merchandiser and its supplier in the history of American business.

The genealogy of the Sunbeam Company was anchored in the Schroeder Headlight and Generator Company, which had been established in 1883 by Charles F. Schroeder. He had slowly developed a headlamp and generator mechanism for locomotives, a reliable artificial sunbeam that was capable of guiding thousands of puffing, chugging locomotives on their journeys through the night. Over time Schroeder's headlamp system found its way through distribution by the Baldwin Locomotive Works to use on the majority of American railroads and in more than fifty countries. Following Schroeder's death, an Indiana corporation was organized in 1918 for the purpose of acquiring the assets of the Schroeder Headlight and Generator Company. The new organization became known a year later as the Sunbeam Electric Manufacturing Company. The Sunbeam Company carried on the traditional products of the Schroeder Company, the locomotive headlight and generator, and simultaneously sought other products to manufacture. Experimenting first with the production of electric plants for farms and later during the 1920s with the manufacture of domestic washing machines, in 1929 Sunbeam entered the electric refrigerator field by producing a refrigerator unit. In a similar fashion, in the marketing of their product, they too secured the outlet of the great Sears, Roebuck Company for their refrigerator unit.[3]

The cover of a Sunbeam Electric Farm-Lite catalog from 1928. The Sunbeam plant was located in Evansville, Indiana. From a Sunbeam catalog in the author's collection.

Over the subsequent years the relationship became an extraordinary alliance between the two manufacturers and the major mass merchandiser. Across

Fig. 1
Sunbeam Farmlite-Model D
Capacity 850 Watts—32 Volts

In 1928, when rural electrification in the United States was geographically very limited, Sunbeam manufactured this gasoline-powered generator that was capable of producing 850 watts of electricity for farmers. From a Sunbeam catalog in the author's Collection.

America Sears stores could be found in major communities and its catalog stores were prominent in smaller locales. In addition the company's familiar catalogs were delivered to households everywhere. From the initial appearance of the first *Coldspot* refrigerator in various Sears stores and later with the inclusion of the *Coldspot* in the 1930 catalog until the outbreak of World War II, the Seeger-Sears association was profitable for both companies. In the depths of the depression of the 1930s, Seeger employed a skilled workforce that could maintain its daily schedule of three shifts, usually five, and sometimes six days a week, to meet its production obligations for *Coldspot* and other refrigerator brands. As a consequence of market demand for residential and commercial refrigeration units throughout the economically troubled 1930s, Seeger became an even more powerful presence in the national refrigeration industry.

Efficiently operating its assembly lines at full capacity, Seeger enjoyed the reputation of being able to deliver to its customers their refrigerator orders on time. Popularly known among its employees as "the house of work," Seeger's management leadership through the years had developed systematic control over essential materials, a network of suppliers, and a corps of employees that staffed every department of operations. If a problem developed in any one of the departments or on the assembly line there was a team effort to successfully solve the matter.

To understand how the Seeger Company had in place the necessary managerial and operational systems to confidently undertake such a significant business relationship with Sears, Roebuck and Company in 1928,

Buyers to the Rescue

Although the following example of how buyers responded to a crisis comes from 1951, it is an ideal illustration of how central the buyers were to the efficiency and cost effectiveness of Seeger's manufacturing operations. On October 4th, the workers for a supplier of domestic wire shelves suddenly went out on strike. Word of the strike reached Seeger on Monday, October 6th. The production schedule was immediately changed to another product that did not require the shelves. The change required advancing the schedule and increasing the quantity of insulation delivered. A buyer immediately contacted the Kansas City supplier for the necessary insulation. By Wednesday, October 8th, the schedule of delivery had been changed to bring the material in two weeks ahead of the original delivery date and in quantities of five carloads a day instead of one. The strike lasted three weeks, but the Arcade Street plant suffered a shut down of only two days due to the strike at the supplier's factory.

Only in a crisis situation such as this were the other Seeger employees even aware of the existence of a problem. In the company records that have survived, there are numerous instances where emergencies arose, but no serious affects were felt by the plant. Deliveries may have been made with but a few hours to spare or substitutions of a different model or product helped offset the loss, but with few exceptions Seeger came through for its customers.

we need to examine the company's professional engineering expertise, management structures, technical capabilities, and systematic assembly procedures. Unlike many of its competitors, in 1928 Seeger had in place the people and processes to do the job.

Engineering Expertise

As early as 1905, the Seeger Company stated in its catalog of residential and commercial products that a staff draftsman produced precision drawings for the construction and materials to be used in all its products.[4] Perhaps it was John A. Seeger himself who undertook the task at that time as he had performed similar services thirty years earlier while leading the workmen installing the woodwork and cabinetry at St. Paul's Assumption Church on Ninth Street. In any event, the preparation of detailed drawings was essential for manufacturing all its products efficiently and at a consistently high level of quality. At a time when there were few widely accepted standards for refrigeration appliances, the Seeger method for making carefully engineered products meant contributed to a high level of buyer satisfaction. As the company's product line expanded over time, it needed more employees who could supply the requisite technical skills. Consequently by the late 1940s, Seeger's Engineering Department had a staff of about 150 persons, which included mechanical, electrical, chemical and ceramic engineers, as well as draftsmen and engineering technicians.

Voices *from the* Plant Floor *at* Seeger

A Financial Windfall for Sears

"'Seeger Management' found that they were making more money than they [had] planned on the refrigerator product line. An analysis pointed toward the high-volume refrigerators sold to Sears. Although Sears' sales had grown substantially, they wanted more of the market. Irving Warren, treasurer of Seeger, analyzed the cost structure and found the bill of materials [on the Sears units] called for two hydrator pans. [He also found that] Sears was billed for two but received only one, a significant difference. The number [of refrigerators] sold to Sears was determined and Walter Seeger and Irving Warren hand delivered a check to [a] Sears officers in Chicago to cover the difference. Integrity!"

Ray Phillips

Voices *from the* Plant Floor *at* Seeger

"A House of Work"

"I was a mechanical draftsman and joined the Seeger Refrigerator Company in 1951. Don Connally and Frank Taylor interviewed me for this position. As I had printing experience I was soon charged with preparing brochures, parts lists, and service lists for the refrigerators and freezers manufactured for Sears, material that required that company's approval before it was printed. In 1951 Seeger has some hospitalization coverage, provided two weeks annual vacation, and sponsored such events as the annual picnic for the employees and their families at Excelsior Amusement Park and a Christmas Party at the St. Paul Auditorium, in the theater section, with such masters of ceremonies as Bob DeHaven and K.C. Jones. Seeger was known as a house of work and Kern Sosey, the Works Manager, ran rough on the factory. I have never regretted working for the company."

John O. Jahnke

Voices *from the* Plant Floor *at* Seeger

Working on Milk Coolers

"As a mechanical engineer I went to work at Seeger in 1952 in the plant engineer's department. Paul Bowman was the head of the department and the best boss I ever had. I learned about the problems of assembling products at Seeger, working on the milk coolers for the Sunset Company. We manufactured 250- , 500- , and 1,000-gallon milk coolers for dairy farmers. Under Paul Bowman's leadership we would work all weekend on solving some kind of engineering problem. Seeger gave us the freedom to do things, to get things done."

Dwain Bell

The members of the company's Engineering Department who were primarily engaged in planning, research, preparation of drawings and specifications, and for the manufacture of refrigerators and freezers were central to Seeger productivity. Among all the technical jobs that this staff handled was the listing and numbering of parts and the making and testing of working models of new products. Their goal was to put a high-quality product on the market that would satisfy and serve the public market. The Engineering Department also had to somehow satisfy the interests of the designers, sales representatives, cost accountants, and production officials. Consequently much of their time was spent in developing practical, innovative, or improved features for Seeger products, through research and experimentation that had not previously been available in the marketplace.

The department was the point of departure for all new or revised products. A first step in this process was a meeting with sales officials and the representatives of the customer for whom the product was to be manufactured. This procedure was followed with both new products and model changes to existing products. Through meetings of this type, the Engineering Department team learned what the customer wanted.

With Sears, Roebuck and Company, Seeger's largest and most significant client after 1930, as well as with Montgomery Ward after 1936, the procedure was different. For these retail and mail-order firms, the engineering requirements for appliances were completed at the Seeger plant in St. Paul and approval with the customer was reached on a cooperative basis. In every instance the design of the cabinet was largely determined within limitations specified by Seeger and by an industrial designer employed by the customer. One of the biggest jobs in the process was the preparation of drawings and specifications. An accurate scale or full-size drawing of each part of each unit had to be made.

A supervisor and a sheet metal worker at Seeger check a blueprint for a sheet metal layout. From a Seeger News *in the author's collection.*

The process by which these drawings for a new product or project were prepared was straightforward and detailed. Preliminary drawings were made by the staff and blueprints repro-

The Central Role of Industrial Designers

As professionally capable as the industrial engineers in the Seeger Engineering Department were in the 1920s and '30s, there were times when they played a secondary role to industrial designers who were hand-picked by Seeger customers such as Sears, Roebuck and Montgomery Ward. For example in 1935 Sears had Raymond Loewy redesign the *Coldspot* refrigerator. Loewy's design, which was marketed as the *Coldspot Super Six,* won design awards and set new sales records for Seeger. Loewy's contribution is covered in detail in chapter 6.

Montgomery Ward hired the Reinecke and Associates industrial design firm to redesign the cabinets for the Ward's line of refrigerators. In the late 1940s Sears turned to Carl Bjorncrantz, the designer of the *Coldspot* freezer, to rejuvenate this product line. These outside industrial designers were responsible for the eye appeal of the cabinets that the Seeger Company built for its clients.

Because eye appeal was considered just as important as reliability and functionality in a refrigerator, the meetings that involved Seeger sales representatives and engineering managers with representatives from Sears or Montgomery Ward and their talented industrial designers were critical to the work that Seeger had to do. Many important changes resulted from these meetings. An example of such a change in the commercial freezer line was the elimination of the 18-foot freezer with a double-door design and the substitution of 19-foot models with a single door. The buying public wanted the convenience of a single door, and through feedback from the customer representatives, this change was made. Once these vital conferences were completed and the basic elements decided upon, the Experimental Department made a sample model by hand. Tests were run on the model and the customer's representatives checked the appearance before final approval was given by both the customer and Seeger. At this point production people were called in so they could submit their ideas on how to produce this new model in quantity. At this point, the detail work in Engineering and elsewhere in the plant began.

duced from them. Early copies were marked in pencil "Not for production." These prints were then circulated to the appropriate departments within the plant for examination and use in obtaining estimates on cost. Where dies were required, the Die Shop received the prints; for the requirements of parts to be purchased, the Purchasing Department was provided with the materials. The parts that took the longest to secure were drawn first so that all preliminary work ended at about the same time.

During the period when the drawings were made, stock lists were prepared showing every part needed on each drawing. Key men, with the assistance of their team, prepared these lists for new models or revised them for model changes. The stock lists showed the part number assigned to every part that would be used, the quantity or size of the material or part, and reference was made to the specification number on the drawing. Duplication by blueprint was then completed. As with the drawings, a distribution list routed the information to the appropriate places within the company. Additional copies of the drawings and stock lists were filed for future reference.

The specifications were detailed descriptions of the material and parts or the processes. Most of these were prepared internally, although some were given to Seeger engineers by the customer. They were listed as "M" specifications, which covered all material and some purchased parts to be used in manufacture and the "E" specifications, which covered all engineering processes. An example of the complexity of these specifications is the documentation for M240-Copper refrigeration tubing. The specifications for this part included exact details on the tubing's chemical composition, wall thickness, tensile properties, expansion, water content limit, and other necessary requirements. Similarly, the "E" series of specifications gave the exact manufacturing process that was to be used to make, for example, E286-Dryout cabinets.

Another important function of the department was the handling of Engineering Change Requests. These forms were completed whenever a change in material, parts, or processes was desired. Before Change Requests even reached Engineering, the Industrial Engineering Department had to first give its approval to them. Of those Change Requests processed by Engineering personnel, about 75 percent were typically approved and the changes requested were made.

Change requests came from three sources. Most of the requests originated in production and other plant departments, while others originated within the Engineering Department itself. The final, and least likely, source was a failure to maintain specifications. A die, for example, that did not turn out pieces to exact measurements could result in the preparation of a Change Request. In this case drawings were revised to fit the actual production part. This type of work consumed a large share of Engineering's time.

As the preliminary drawings were completed, additional models were made by the Experimental Department for further testing both at the plant and in the field. Changes were made as discrepancies were discovered. When the drawings were acceptable and complete, a check sample model

Voices *from the* Plant Floor *at* Seeger

One Spouse's Memories of Seeger

"My husband, Philip Turner, a St. Paul native and graduate of Humboldt High School, was an employee of Seeger and Whirlpool from 1952 until his death in 1979. He was a design draftsman and held several patents for parts that he designed for appliances. He loved his job and was a very loyal employee. The annual company picnic at Excelsior Amusement Park was a wonderful family day. Our daughter worked on the assembly line during the summer earning money for her college education and our son received a scholarship for Outward Bound, paid for by Seeger."

Mary Turner Maher

Voices *from the* Plant Floor *at* Seeger

One Senior Engineer's Recollections

"I am a chemical engineer and joined Seeger in 1938 working as a lab chemist in quality control. Seeger was classified by some persons as 'slave driving,' but for engineers it was an exciting place to work. On Fridays we would say that the week was half over! We were given lots of rope to run with. We were encouraged by the management, and we were not berated or beaten up if we made mistakes. Kern Sosey, the Works Manger, never wore a watch. He didn't want to know the time. I climbed the ranks at Seeger, serving as Director of Manufacturing Engineering, retiring in 1973 after a thirty-five year association [with the company]."

Paul Bowman

was made that gave the Engineering Department staff another opportunity to learn of any problems or quality issues. This whole process, however, could take from three to six months.

Experimental and Developmental Department

When the Engineering Department had brought a project to an end stage, then the Experimental Department became a tool for the engineers. At Seeger the manufacturing departments looked to the Experimental Department for producing and assembling new products. The greatest help generally came, however, from people outside the department itself—the employees on gang presses, the Metal Assembly Lines, and other departments. These workers lived with the job. They knew the strengths and weaknesses of each part and assembly. They had first-hand experience with a particular job every day, and when problems arose, they might think to themselves: "Why was the part designed this way?" or "Why was the assembly undertaken in this manner?" or "Would the process be more efficient if it were done in some other manner?" To encourage on-the-job criticisms and suggestions, Seeger established a "Suggestion Plans?" program, seeking new ideas for the production of more competitive refrigerators and freezers. The task at hand was not limited to a few engineers; it required the efforts of the whole organization.

The Experimental and Development Department was self-contained so that it could handle an idea from the drafting board to the finished product. For example: an idea conceived by anyone, that also had merit, passed through the following process. First the idea was recorded. Then a group of technical staffers evaluated it and, if appropriate, tried out the a procedural or process change or improvement on department equipment. The department functioned like a production department, working under a fixed schedule and from regular engineering production drawings. During the hand making of a new part or model, the manufacturing departments were consulted and their approval was obtained before releasing final drawings.

After the part or model was made and reviewed, the item was tested for physical and chemical characteristics, and under set ups, destructive tests were made in addition to the vibration, impact, and drop tests. New products were under development and required close coordination among the Experimental and Engineering Departments and the production shop.

The next step was to undertake the tooling program for the product. Dies were ordered, purchased parts were scheduled for delivery, and jigs were completed. This process also required several months. When everything

A mechanist adjusts a hand-operated hydraulic press for stamping sheet metal parts. From a Seeger News *in the author's collection.*

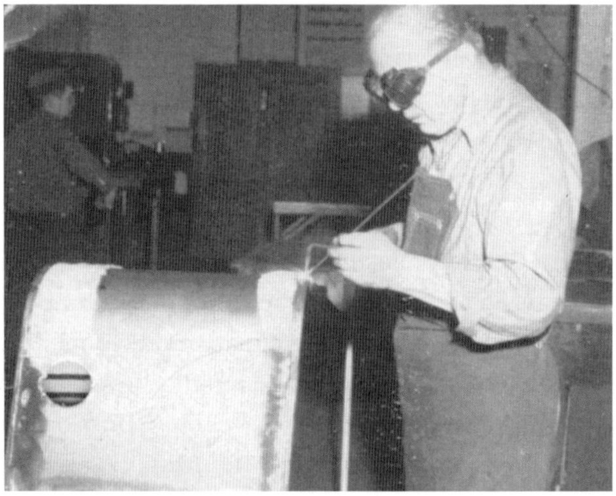

A welder finishes up some specialty work on a new refrigerator part. From a Seeger News *in the author's collection.*

seemed to be in order, the "go-ahead" signal was given for a production pilot run, which normally consisted of manufacturing twenty-five sample units.

To test a new product, the pilot run was generally scheduled six weeks ahead of production. Samples of die-made parts were carefully checked. Changes could then be made before or after the pilot run. This run gave production the opportunity to discover any bugs and to properly organize the production procedure.

With the pilot run complete and the model scheduled for production, the Engineering Department stepped to the background but not out of the picture. It continued to control any changes that affected the department's contributions to a product either directly or indirectly. In all cases, Engineering had to approve all engineering change requests. Throughout the production period tests were run on all models. When each unit of production was to be checked, the Inspection Department handled the work on the basis of specifications set up by Engineering.

More intensive sample testing was done in two test rooms, one for freezers and one for refrigerators. About one percent of production, two freezers and four refrigerators per shift, were checked each day in a sixteen-hour series of tests that covered every phase of operation. In this way when trouble was discovered, corrective action was taken quickly.

In addition to these test rooms other controls were maintained. For example, a moisture test was conducted to insure that the refrigerant contained an average of less than five parts of moisture per million parts of refrigerant. Porcelain was tested for adherence to the metal. The paint application was tested both by the Engineering Department and the DuPont

Company, which supplied the paint that Seeger used. The Engineering Department cooperated with the Traffic Department in conducting shaker and impact tests, in working out requirements for safe transit from the plant to the customer's warehouse, and in discovering better and lower cost packaging for Seeger products.

Industrial Engineering Department

Industrial engineers, who observed an employee's every movement to determine the normal time required to do a particular task or job, helped improve manufacturing operations and maintain efficient productivity in the Seeger factory.[5] These engineers carried out time studies. The times they recorded from actual operations were used to determine a "rate"—if the rate was "tight," it was unfair to Seeger employees and to the management. First of all, a tight rate was unfair to the employees because they were not receiving their just due for their effort and contribution to the total enterprise that resulted in the sale to the customer. Secondly it was unfair to the management because it could result in a false labor cost and could result in labor unrest.

If the rate was considered "loose," it was also unfair to the employees and to the management. First to the employees because they received more than their just due, which ultimately reflected in an excessive sales price to the customer. The customer would probably refuse to buy the Seeger product and this could result, ultimately, in loss of work for the employees. Secondly it was unfair to the management because the result could be a loss of business and contribute labor unrest.

The tightrope in determining an accurate rate was a precarious one and required close attention to details to determine what was right and just. The "rate setter" was a representative of the Time Study group of the Industrial Engineering Department. Although that group was the one with whom an employee had most frequent contact, Time Study was only a single phase of the department's work. The other major functions were Methods, Estimating and Billing, and Clerical.

The department's Time Study section and its file of 18,000 group and individual rates had to be constantly revised and brought up to date. Each new product or model required additional studies and rates. To carry on this work, staff members were assigned an area of operations such as the

An industrial engineer (left) conducts a time study of assembly procedures on the Commercial Assembly Line. From a Seeger News *in the author's collection.*

Voices *from the* **Plant Floor** *at* **Seeger**

Competition between Shift Units

"I joined Seeger in January 1941, working on the second shift between 3:30 P.M. and midnight, as a timekeeper on the domestic lines [that were] manufacturing refrigerators for Montgomery Ward and Sears. There was production competition between the shift's 500 unit versus its 510 unit [these numerical names came from the assembly line designations] [in the number of refrigerators we could assemble in a shift].

"Sometimes I would sleep in the lobby on Friday nights to be ready to work the overtime shift on Saturday from 7 in the morning until 3:30 in the afternoon. Later I moved to the industrial engineering department setting price and time on products. My first assignment was in the wood mill, as Seeger was still making walk-in units for butcher shops. [I also] worked for a fellow named Dan Eden, and later for John Schwenke. Bob Ahrens was the chief engineer for the company and a great fellow.

"After World War II John [W.] Seeger was head of industrial engineering. It was an exciting time—we had forklifts, pallets, and experienced major changes in the process of manufacturing our products. It required a new layout for storage facilities. Some of the older employees resisted the changes, but they had to make the transition. The hourly employees were terrific—they offered good suggestions for improving operations. Problems were everybody's problem. I never asked for a raise. It was always there when I thought I had earned it."

Richard Hagberg

Assembly Lines, the Metal Assembly, and the Metal Shop.

Much time was spent discussing time studies, but few employees were familiar with the procedure used in taking the study and what was included in the rates. A new operation or a revision in an old one made it necessary for the section of the factory to schedule a study. When the evaluator from Time Study arrived at the site of the process or procedure that was to be examined, his first act was to generally observe the job. He checked to see if the worker's actions were smooth and economical, in other words, that there were no wasted motions. The position of the materials or tools to be used before and after the operation might have an important relationship to the operator's tasks. After this initial observation, the operator might suggest minor changes to improve the flow of work. This done, the actual study began.

The job or process was then broken down into elements or motions. A simple example would be to pick up a piece and place it in a particular machine. The second element or motion might be setting the machine into motion by means of a button or foot lever. The third would be the removal and placing of the completed part in a nearby location for subsequent transfer to another area of the plant. Each motion had to be broken down to the point where it would give an accurate description of the job but not to the point where it could not be accurately timed.

When all the elements had been identified, the timing began. Seeger staff used the continuous study method for timing. This involved starting a stopwatch at the beginning of the task and allowing it to run throughout the study. Each time the piece was picked up and placed in a machine, a record was made of the length of time taken to do this. This motion may have been made a dozen times during the study. An average elemental time was then calculated. Special conditions that arose during the study were noted separately.

Throughout the study the operator was rated. The rating was based on what an average operator could do under normal conditions without undue fatigue. The effort rating for each element was applied to the average time to determine the normal time. The total normal time was used to determine the rate. With this information converted to production figures, industrial engineers could then determine the number of employees of various skill levels that a particular department would need under normal working conditions and also forecast schedules for each operation that had to be completed to meet customer requirements.

Although the major portion of the Time Study group's work was concerned with operational studies, they also were directly involved in methods and work simplification. Their part in this work generally took place just before a study was taken or in cases where the changes were relatively minor to the overall operations. Major changes in method would normally be turned over to the Methods groups.

Additional duties of the group were to estimate the Assembly, Dulux, and Porcelain labor costs for new products, new models, and revisions on existing models. Normally Time Study personnel worked out these costs by visualizing all the operations involved and then developing their estimates by using current similar rates or by estimating new operation rates. Still another job that they had responsibility for was the determination of Standard Labor Cost, the direct labor cost for each model.

Methods Group

A less familiar, but equally important section, of the Industrial Engineering Department was the Methods group. This group dealt with domestic and commercial products by determining a department's floor layout, all the operations a department was expected to perform, and the proper sequence in which these operations were to be conducted in the manufacture of specific products. In simple terms, they determined the methods to be used to do the work.

More specifically, their duties began with the placement of the workers and machines for the most economical and efficient production for a given quantity of parts or for a specific operation that was part of a larger operation. This included placing machines around the shop floor so that adequate space was available for needed raw materials and for the temporary holding of finished parts. The worker's position was, of course, determined by the operation on which he worked. The Methods group determined the sequence or order of operations. Such layouts were made for both new products and revisions made necessary by changes in current products.

One of the most important duties was the final routing on parts and assemblies for all models. Master routers or diagrams were prepared showing each successive operation required to process each part. One of these routers existed for each part fabricated. From the routers, the Methods group prepared instruction cards that gave complete details for the performance of each operation. Thus when the Metal Shop was ready to begin parts production, it requested these cards and the associated information needed to do a specific manufacturing step.

Less complex, but nonetheless important, was the job of investigating Engineering Change Requests (E.C.R.). Each proposed change in an operation required an E.C.R. When the Methods group decided a change was acceptable, it then had to revise all the requirements and specifications that were affected. All the routers were then changed. Many of suggested changes resulted from the work simplification program that was constantly carried on by the industrial engineers of both groups. Work simplification referred to the process of discovering easier ways to do the work. It was a major responsibility of the Methods group. One of the group's most valuable contributions was to assist production departments in working out problems on new processes and new equipment. The engineers in the Methods group frequently had the experience and knowledge to quickly overcome such problems.

Finally members of the Methods group also prepared cost analyses to justify the installation of new equipment and plant changes. Typically these analyses showed that by incurring an expense now the company would receive a future savings in labor or material or both. Overall the Methods group's duties added up to a big job for a small number of engineers and support staff.

The highly trained personnel of the Industrial Engineering Department also prepared cost estimates for new products and new models. Typically their concern was with the Shop and Metal Assembly Departments. The Time Study group would prepare rates and cost figures based on their studies. Normally the estimates they developed for both the Shop and Metal Assembly Departments proved to be remarkably accurate. Typists in Industrial Engineering then turned out voluminous reports and other materials for use by managers throughout the company. Given all the research and analysis they provided to these managers, the Industrial Engineering Department could be considered a research organization. Although it was largely absorbed in the detail work of studies and routers, it always had one eye on the future. Each member of the department was constantly on the lookout for better and faster ways to reach the common goal: high quality production.

Purchasing Supplies for Production

Acquiring essential raw materials and critical parts for the production of Seeger refrigerators was a key management assignment.[6] On the surface, purchasing supplies appears to be a simple task. Make a few calls to vendors, contact the appropriate person in sales, and then check the items off on the order requisitions, just like buying the items on a supermarket list. At Seeger hundreds of firms sought to supply parts for the manufacturing process. Calls came in from these potential vendors in significant numbers, and they were often supplemented by telegrams and letters. The sales representatives of all sorts of material, products, and supplies were continually seeking an opportunity to put in their bids and to sell their products. Instead of a dozen or twenty items, the buyer's list at Seeger typically contained some 7,500 items that might need to be supplied at any given time.

Purchasing, however, faced not only one deadline when a specific product needed to be ready to ship, but also many deadlines every day of the year when individual parts or materials needed to be in the plant ready for use on the manufacturing floor. Frequently conditions or situations would come up that changed the entire schedule for certain deliveries. Consequently the deadline or the order or both might have to be immediately changed as a result. Appearances in the Purchasing Department were deceiving. In fact the department managed an extremely complex process.

The Purchasing Department was divided into two sections, Purchasing and Records Control. In the Purchasing section staff members included buyers who were responsible for the purchasing all commercial cabinet parts, electrical goods, rubber goods, copper fabricating parts, lumber, fittings, production printing, and laminated plastics. Other members of the department were assigned to purchase plastics, hardware, screw machine parts, stampings and forgings, while another employee's duties included purchasing all maintenance supplies, office equipment, printing and supplies for Engineering, Jig Shop, and the Laboratories departments. In addition a different staff member bought a wide variety of items such as crating, insulation, wire shelves, corrugated board and cartons, Masonite fiberboard, fasteners of all types, paper, metal plating, and other types of outside services.

The Purchasing Department worked hand-in-glove with the Engineering Department. As the engineers developed new or revised refrigerator models, the preliminary or "not-for-production" blueprints of parts that had to be purchased were sent to the buyers. They were then responsible for obtaining price quotations. By means of telephone, mail, and personal contacts, bids were obtained from at least three to five possible suppliers for each item. The exceptions to this would be a patented item or one that

The conveyor on the shipping dock took bags of insulation that had just been unloaded from a railroad boxcar inside the plant for storage until needed. From a Seeger News *in the author's collection.*

required expensive preparation or tooling. These items would normally be obtained from only one supplier.

Securing several bids served a twofold purpose. It assured the Seeger Company of the lowest price and also of having a secondary source of supply in case the primary source failed to deliver. The prices so obtained were then filed until needed for the orders that would go out later when production was underway.

Available records and interviews with long-time Seeger employees indicate that the bulk of all purchases were made east of the Mississippi River. These parts and supplies arrived from points as distant as Massachusetts, Maryland, and Delaware on the East Coast, and California and Washington on the West Coast. Steel was bought mainly from the Chicago area, but some also arrived from Pittsburgh and Granite City in Illinois.

Each year about 1,500 railroad carloads of material arrived at the Seeger loading dock. Additional material also arrived at Seeger by truck and in special cases by plane. All of these deliveries needed to be recorded, checked, and the various items supplied unloaded and delivered to various storage areas around the plant. In the 1940s and '50s Seeger had the distinction of being the largest user of copper tubing in refrigeration grades in the United States. Approximately a million pounds of aluminum arrived each month, the greatest quantity of any manufacturer in the Upper Midwest.

The Records Control section of the Purchasing Department was responsible for the tedious and time-consuming record keeping and to improve the efficiency of order handling. This section made every effort to standardize its procedures and to insure the on-time delivery of the materials the buyers had ordered. The Purchasing Department's role in production began with this section.

As we have seen, the Engineering Department, along with the Experimental Department and Engineering Laboratory, first had to build and test a model of a product, a refrigerator, or freezer, until it was satisfactorily completed. Simultaneously, the company's Sales group worked with the representatives of a specific customer, say, Sears, Roebuck and Company, or Montgomery Ward, or Frigidare to define all delivery and price arrangements for the refrigerator or freezer that was under discussion. When Engineering, Sales, and the customer were in agreement, final approval was given for the model or models designed. This approval meant the acceptance of raw materials, parts lists, and all other specifications for the product. When these steps had been completed, all was in readiness for the first order.

The customer then placed an order, usually three months or more before delivery was expected. The Sales group, who worked with the customer, and the Works Manager at the factory, who represented production and purchasing, then signed off on the release for production. These sign-offs insured the coordination of all parties involved so production could begin and delivery would be made by the dates arranged. A release schedule was then prepared to coincide with the delivery dates. A given order, called a job order, might call for 5,000 or 10,000 units to be produced. A number was assigned to the job order. The quantities to be produced and dates for delivery of the orders would be determined in conformity with the anticipated needs of the customer.

Bins such as these held just some of the thousands of small hardware items that had to be on hand for manufacturing refrigerators and all the other products that Seeger made. From a Seeger News *in the author's collection.*

The resulting release schedule was sent to the Stores Department. This department, in turn, ordered two copies of the stock lists for the model or products to be manufactured from Engineering. For each item on the stock list a travel card was prepared. This card indicated the number or quantity of each item needed and the date by which it had to be shipped by the vendor or source. The shipping date was normally set five weeks before the material was needed in the Seeger plant.

The travel card, along with one copy of the stock list, was sent to the Records Control section.

The receipt of the travel cards was then entered into two separate records. One record noted the date on which the travel card was received and the other recorded the job order number. It was also checked to see that each item had been ordered. The travel cards and stock list were then turned over to the appropriate buyer.

When the buyer received the travel card, this employee attached the price record file to the travel card and would select from the file of cost estimates or bids the source from which the purchase would be made. While preparing the order, the buyer had to decide how much or how many of an item could be stored on site. Inventories of items carried in stock varied from one to three weeks. Bulky items such as insulation and crating were brought in almost as used. Small items such as screws or nuts would be purchased for the entire job order.

The Chief Purchasing Agent then approved the outside source or vendor that the buyer had earlier selected or identified. This approval was very important because the Chief Purchasing Agent might have been aware of special conditions relating to this purchase that were not known to the buyer. The buyer had seven days to complete this step.

The original of an order was then sent to the supplier. Another form, an acknowledgment, was included with the order. The supplier was asked to return this acknowledgment as soon as he was certain he could fill the order. The name of the supplier and date of the order were then placed on the travel card. This card with the remaining copies of the order was returned to the Records Control section. Appropriate entries were made in the travel card record books to show that the order had been placed. Records clerks prepared a file folder for each order. Into this folder would go the remaining copies of the orders, which would be available for use in the event a follow-up was needed or any other correspondence was generated pertaining to a particular order. The folder was then filed alphabetically. Each clerk was assigned a portion of the file and was made responsible for the follow-up on the orders therein.

Records Control clerks regularly checked the travel card record books. If they discovered an entry for which the travel card had not been returned within the seven days allowed the buyer, this fact was reviewed and an attempt was made to determine why the card had not been returned.

The Records clerks then followed each order until it was acknowledged or delivered. This required mailing additional copies of the order, or sending out tracers, and in cases of urgency sending telegrams. If the acknowledgment was not received soon after mailing, the supplier was contacted to find out why the acknowledgment had not arrived. This constant follow-up helped insure delivery on time.

Normally the acknowledgment would be received in due course and delivery was made on or near the date intended. A copy of the original order was sent to the Stores Department. When the initial set was completed, this form was used to reproduce the information on the receiving report when the material arrived. A comparison was made with the order information to discover any errors in quantity or type of material delivered. The material then passed to Receiving Inspection where it was carefully checked. If the material met specifications, a copy of the receiving report was sent to the Records Control section. Any order might contain rejects. If a trend in the number or type of rejects was noted, the Purchasing Department buyer was given a report. The buyer would then contact the supplier to arrange for correction of the defects. In cases where the material failed to meet specifications, the buyer was also notified. Upon receipt of this notification, the buyer then determined whether to accept the material as is or to arrange for immediate replacement.

When the Records Control section received the delivery report from the Stores Department, it was filed with the original order. If the order was complete, the folder was then placed in the inactive file. In many cases a particular shipment would be only a partial one. When this occurred, the folder remained active until the entire quantity had been delivered. Copies of the receiving report and all invoices were checked for agreement with the order. They were then sent to the Accounting Department for payment. Purchasing so many different parts and materials from so many sources invariably produced some problems. The most serious was, of course, suppliers who failed to deliver on time, many through no direct fault of their own. Equipment breakdowns could rob them of valuable production time. Flat wheels on a railroad car might sidetrack a load of material that was vitally needed. Sudden washouts on a rail line, winter snowstorms, floods, or icy roads could have delayed a train or truckload of material. Convincing the operational managers of railroads or trucking lines that a delay on the delivery of the parts or materials that their railcar or truck carried would affect some 2,000 people at the Seeger plant was no small task. In addition to these delays caused by equipment failures or severe weather, sudden strikes could cut off a vitally needed part completely. A strike could be particularly serious when it affected the delivery of bulky items such as crating and insulation that could be maintained only on a two- or three-day level of inventory

Given all the parts and materials that the Seeger plant needed at any given time, effective managerial coordination among the many departments involved in engineering and purchasing was critical. In addition the long-term success of the Seeger operations hinged on the willingness of all the employees involved in these complex procures to follow consistently the necessary steps that were involved in their departmental duties. Because those in charge and those who did the work in the plant over time were able to execute these procedures across a wide range of refrigeration products, the Seeger Company became a national leader in refrigerator manufacture.

5

Parts and Tools for Production

A MANUFACTURING FACILITY RESPONSIBLE for assembling tens of thousands of refrigerators must have the capability to maintain control over an adequate inventory of supplies at all times to accomplish the task.[1] The Store Room was the department of Seeger that supervised the inventory and supplied production operations. It handled all the steel and prefabricated parts and materials that were used in the manufacture of the residential and commercial refrigerators by the firm. Department staff members received and unloaded more than the six thousand items needed to maintain production schedules, checked them, stored them, issued them, and delivered them at the proper time to the correct place. It was an essential and comprehensive operation and each member of the department was an important part of supplying the producing and assembling departments of the company.

The accompanying map (Figure 1) illustrates the distances over which the various items traveled to reach St. Paul and the plant. The boundaries of the supply area extended as far east as Massachusetts, as far south as Louisiana, west into California, and north to Crosby-Ironton, in Minnesota. Within those boundaries were a hundred individual companies that shipped materials to Seeger. All methods of transportation, train, truck, barge, and plane brought the materials to St. Paul. These supplies, totaling approximately six thousand items, were then stored in the appropriate one of the company's several warehouses and buildings.

Railroad was the most common transport method at Seeger. On any given day, a many freight cars stood adjacent to the plant in the area below and to the east and west of the Arcade Street Bridge. With few exceptions, each of these cars was loaded with materials destined for Seeger or one of its corporate neighbors. Rail shipment was particularly necessary for large or bulky materials. Typically Seeger suppliers would send the company advance notice that certain materials had been shipped. Each morning a clerk in the Store Room telephoned the railroad that the shipper had used to determine what cars had arrived on that line. With this information a certain amount of

Figure 1

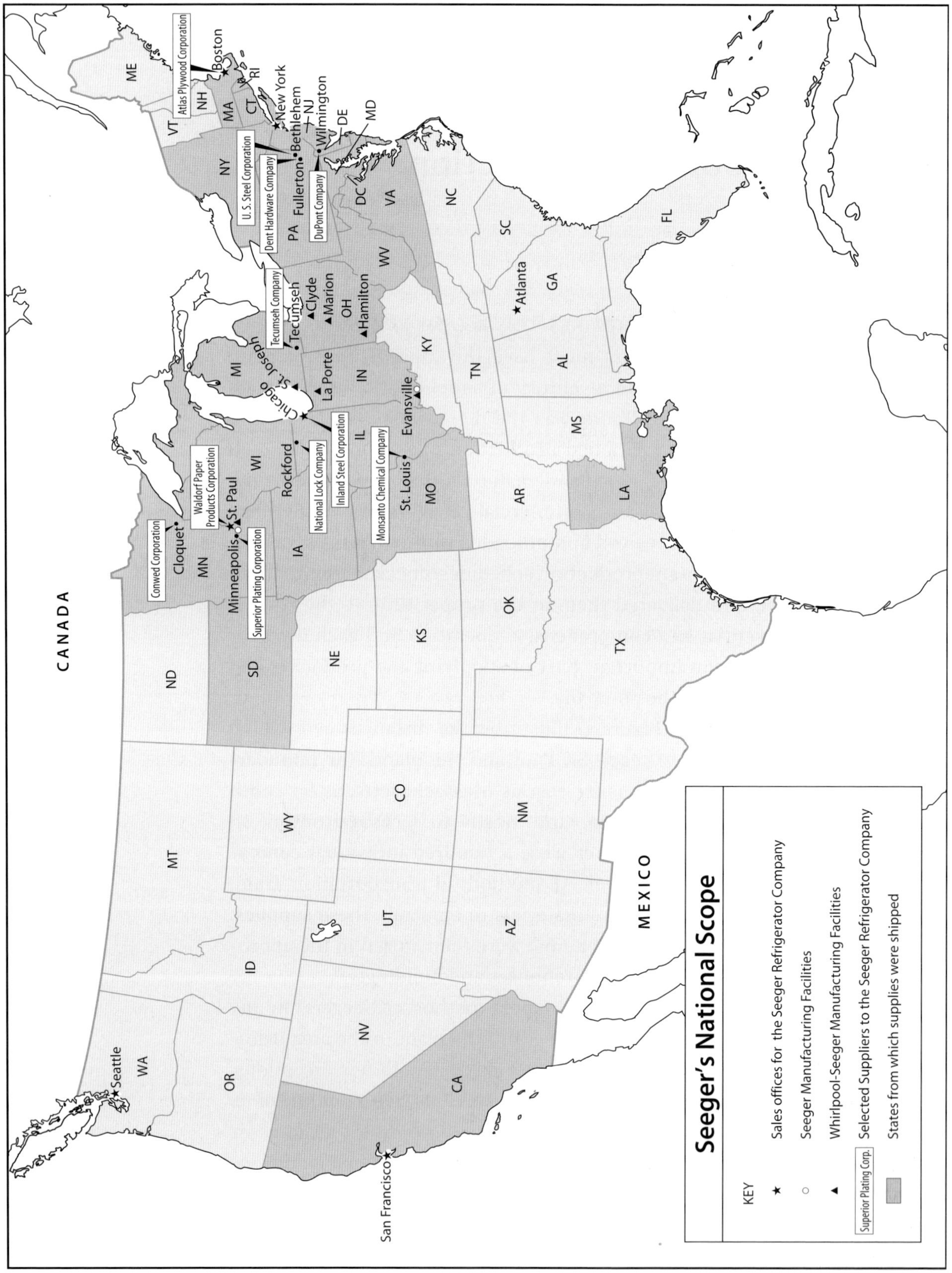
Seeger's National Scope
KEY
Sales offices for the Seeger Refrigerator Company
Seeger Manufacturing Facilities
Whirlpool-Seeger Manufacturing Facilities
Superior Plating Corp. Selected Suppliers to the Seeger Refrigerator Company
States from which supplies were shipped
CANADA
MEXICO
Seattle
San Francisco
Atlanta
Boston
New York
Bethlehem
Fullerton
Wilmington
Chicago
St. Joseph
La Porte
Tecumseh
Clyde
Marion
Hamilton
Evansville
St. Louis
Rockford
Cloquet
Minneapolis
St. Paul
Atlas Plywood Corporation
U. S. Steel Corporation
Dent Hardware Company
DuPont Company
Tecumseh Company
Inland Steel Corporation
National Lock Company
Monsanto Chemical Company
Waldorf Paper Products Corporation
Conwed Corporation
Superior Plating Corporation
WA
OR
CA
NV
ID
MT
WY
UT
AZ
CO
NM
ND
SD
NE
KS
OK
TX
MN
IA
MO
AR
LA
WI
IL
MI
IN
OH
KY
TN
MS
AL
GA
FL
SC
NC
VA
WV
PA
NY
VT
NH
ME
MA
CT
RI
NJ
DE
MD
DC

advance preparation was possible to insure more efficient and expeditious handling of the material. The Store Room crews handled from 300 to 375 freight cars each month. Each working day railcars from the various roads that served St. Paul were switched onto the four lead-in tracks at the Arcade Street plant: at midnight, in the morning, at noon, and at 4:15 PM. The cars then had to be unloaded as they arrived and the weekend accumulation always required a great deal of hustle on the part of the unloading crews.

This truck just delivered wood crating materials (stacked far left) that were used in shipping finished products. The Seeger plant had its own rail spur for the delivery of the many carloads of raw materials and supplies needed daily. From a Seeger News *in the author's collection.*

After rail delivery, delivery by trucks was next in order of the quantity of material handled. No advance notice was received of their arrival so they had to be unloaded without benefit of preparation. Trucks usually handled the smaller and less bulky items that the Seeger operations needed. From 200 to 250 trucks delivered material each week. In most cases these trucks, as well as the railroad cars, carried materials directly from the manufacturers. In addition to the usual truck deliveries, there were approximately two hundred parcels that arrived each week by special and the regular mail deliveries. These were usually small items that were used for miscellaneous jobs rather than on production. A designated member of the Store Room accepted delivery of these packages on the receiving dock and then made sure they were routed directly to the appropriate departments.

The third type of transportation was river barge. Barge transportation was used extensively during the summer season to haul units for the Sears products. It was restricted to the seasons during which the Mississippi was navigable. For certain Seeger products, barge transportation provided a considerable cost savings over railroad hauling.

Seeger managers employed airplane express delivery with some frequency, but it was restricted in the quantity of material carried. Consequently air express was used in cases of emergency, for example when manufacturers were unable to meet

Voices *from the* **Plant Floor** *at* **Seeger**

Storing and Shipping Finished Refrigerators

Seeger did not have the space to store its products as they came off the assembly line. Refrigerators and freezers were stored around town at the Central Warehouses and the facilities of the Murphy Company. The appliances were moved by every type of transportation. The Murphy Trucking Company had a long-standing contract with Seeger for over-the-road hauling, while in the spring and summer months the refrigerators manufactured for Montgomery Ward were sent by barge to Davenport, Iowa, and then by railroad to Dayton, Ohio, where Frigidare compressors were installed in the appliances. The company had a Chicago & North Western Railroad siding at the factory and its products were shipped on every railroad line.

Gene Gardner

Seeger production schedules required the constant use of forklift trucks such as this one to move supplies, materials, and finished goods in the plant. From a *Seeger News* in the author's collection.

production schedules and certain parts were required for immediate use in the plant. Almost any size item could be carried by this method, but items needed in production quantities had to be small and relatively light.

The division of the Store Room that handled the heaviest material was the steel issue. The capacity of the warehouse at the Seeger plant was 5,600 tons. Included in the inventory on hand were 350 sizes and gauges of steel. Examples of the types held in stock were cold rolled steel, hot rolled steel, enameling iron, galvanized iron, copper, aluminum, and bar stock. Each of these, of course, came in various sizes and gauges because fourteen or fifteen sizes and gauges were used in each refrigerator cabinet that was produced. The bulk of a refrigerator cabinet was made from twenty-gauge steel, but the range was from ten- to twenty-eight gauge.

Until just after World War II, a large crew and a great deal of backbreaking labor was essential for the handling the large quantities of steel received each day. As the railcars were shunted onto the siding alongside the Seeger plant and the trucks pulled into the loading dock, this crew unloaded the materials received onto hand trucks with the help of a roller type of conveyor. This work had to be done outside in all kinds of weather, including rain, sunshine, snow, and below-zero temperatures. Then the hand trucks that were loaded with steel were carefully wheeled to the proper fabrication departments. In those days it took five men eight hours to unload a railroad car containing the usual sixty tons of steel. In other words, a full shift was required for five men to unload one car.

The Yale overhead crane in the Seeger plant was indispensable for unloading deliveries of raw steel. From a *Seeger News* in the author's collection.

In 1948 it took only four men one hour to unload that same car containing sixty tons of steel. The installation of a Yale overhead crane in the steel storage warehouse made such a dramatic reduction in receiving time possible. This crane with its ten-ton capacity was capable of easily and quickly handling the five-ton bundles of steel. These bundles were then piled to the rear of the warehouse in stacks, ten to fifteen bundles high.

The process of issuing the steel was as easily done

as that of receiving it. An area in front of the stacks was kept open for the distribution of the daily requirements. Each morning the steel issue section got a shearing schedule, which spelled out the plant's steel requirements for that day. The foreman of the shearing section issued additional information or instructions regarding the exact order in which the various types of steel would be needed. With this information, a crane operator then used the Yale lift to pick out, hoist, transport, and lower the appropriate steel bundle in the distribution area. The bundle was then broken and the required sheets of steel separated and loaded onto hand trucks, then took the steel where it was needed. All of this work was done in the warehouse, rather than on the dock, out of the extremes of the weather and a fourteen-man crew was all that was required for the receiving and issue of the steel.

There were, of course, many more items needed before a finished cabinet could be shipped to Seeger customers. The steel, although heavy in weight and large in bulk, accounted for only a relatively small part of the Store Room's activities. Other than steel, prefabricated parts and materials accounted for over 6,000 items in the inventory of supplies. This inventory of supplies ranged from the large, bulky items such as insulation, counter and shelf glass, crating materials and units, to smaller more compact items as screws, nails, and hinges. For about one day's production of 1,000 refrigerators for the domestic market, four cars of insulation and two cars of crating had to be received and delivered to the domestic assembly line.

The Store Room was divided into two sections, receiving and issue, each employing about fifty men, for handling these other supplies. To better understand how production got these supplies, let's briefly follow the procedure by which incoming material was handled and then its counterpart, issued.

As supplies rolled into the various docks, crews were assigned to unload the material and haul it to the receiving bay. A checker from the Store Room made a preliminary inspection of the packaging lists to determine whether they correspond in quantity and parts number to the freight bill. Any overage, shortage, or damage that the checker noted was merely reported. The Store Room staff member who did the checking made no attempt at adjustment at that time. The shipper's freight bill was always signed. From this initial inspection, the checker made up a receiving card in four copies; one for the Receiving Department, one for Receiving Inspection, and two for the Purchasing Department. Each copy contained the name of the type of material received and a number that was assigned to each type of material in the shipment. One of the Purchasing Department cards was a "flash" copy that was sent to that department within two hours of receipt of the material with only the information obtained from the initial inspection.

After a shipment was unloaded, it was inspected in the Receiving Department before being warehoused. From a Seeger News *in the author's collection.*

The other copies were not completed until a more thorough inspection had been made. Each package in the shipment was marked with the assigned receiving card number.

The Inspection Department then made a more thorough and accurate check on size and defects of the materials received and matched this information against the appropriate specifications and prints. From the information thus obtained, the other cards were completed. The location of the material was noted on the Store Room copy of the receiving cards. Rejected material was stamped as such on the cards and the material was sent to a special bay or area reserved for this type of material. Inspection then made a written report giving the reason for the rejection, which was forwarded to the Purchasing Department. A knowledgeable member of the Purchasing Department subsequently determined whether the material may or may not be used, depending largely on the type and significance of the defects and the availability of replacements. Each type of material had its own assigned area. In the years when the Seeger plant had its greatest diversity of products, the Store Room took in all or part of fourteen different buildings, which gave it a total of 109,520 square feet of floor space.

Probably the most difficult part of inventory control for production was to issue all the various materials in the correct quantities and at the same time keep everyone satisfied. Simply because the production schedule called for a certain number of refrigerators to be produced by a certain deadline, this production order did not mean that all the necessary parts had been received. Realistically Seeger personnel could not stockpile all the items

This photo shows just how big the steel storage warehouse was in the Seeger plant. From a Seeger News *in the author's collection.*

they would need in advance. Many suppliers were behind on orders or at best, just able to keep pace with the orders. Insulation is a good example of just how tight the schedule was. This material was frequently received on the morning of the day it was to be used.

Two methods were used to issue of material. The first was by material requisitions and the second by the bill of material. Either departments that were not directly involved in production or production departments themselves initiated material requisitions for items not directly involved in production. These requisitions were also used for items that had been rejected because they were broken or damaged in transit or in use or that failed to fit. Such requisitions were sent to Store Room where the supplies were subtracted from the inventory records.

The Store Room Department initiated the second method, the bill of material. When the future production schedule was laid out, the Store Room Department requested a stock list from the Engineering Department. This was simply a list of the parts needed for the particular type of cabinet scheduled. A list was made of all parts that were purchased outside and Purchase Orders were made for requirements not in stock. When the production schedule was issued, Store Room requested another copy of the stock list. This time the quantities were listed in the succession the material is needed. For example, if the production order called for 10,000 units to be scheduled in runs of 5,000 units during one week and a week later another 5,000, the parts would be scheduled to fit the runs rather than the whole production order. The items were then pre-posted to determine whether a shortage existed in any of the required items. Parts that had been ordered but had not arrived were noted as to their method of arrival via railroad, truck, or some other means.

Hardware issue men loaded the required parts on skids or pallets in the quantities listed and for the production groups by whom they would be used. Once the parts were on pallets, they were kept in the Store Room until the day before they were scheduled to be used. A material control man was responsible for seeing that the material reached the proper place in the plant at the proper time.

Panel Room items, items that were too bulky to deliver in full-run quantities, were issued daily. The number of such items that had been issued was reported to Store Room so they could be deducted from inventory records as they were delivered each day. This included material such as crating, shelves, and so forth.

Replacements for rejected material were issued on material requisitions as mentioned above. Inspectors reported on such material and sorted it into line rejects, material that had been damaged in use, manufacturers' rejects, and material that proved unsuitable for use. Such replacement requisitions were made out in red. These requisitions had to be approved by the department supervisor and the works superintendent. Store Room deducted the approved requisitions from the inventory books and a Store Room clerk then filled the requisitions as indicated on the forms.

For the most part the material that had been requisitioned was delivered internally on hand trucks or by hand-lift dollies (small dollies that used a hand-operated lever that activated hydraulic cylinders to raise pallets off the floor) that a worker could use to wheel the material through the plant. There was, however, some self-propelled equipment that could be used on the heavier material. These machines included a Yale forklift truck with a 2,000-pound capacity, two electric Yale transporters, and a portable elevator used to handle large units. With the aid of this equipment, the parts and materials were available at the proper time and place with few exceptions.

It was a big job and it was a job well done, serving fifteen departments of Seeger, and all these departments informing the Store Room staff of their needs and while seeking to meet production schedules. Store Room personnel had to chose from the thousands of items in inventory the proper ones and deliver them exactly where and when needed. This support function required constant supervision to maintain its efficiency and accuracy and that meant lots of extra hours of work, particularly when a critical item failed to arrive as scheduled.

Metal and Tool Fabrication

All of the products that the Seeger Company manufactured during its long history required the use of steel or some other metal.[2] These materials had to be shaped and fabricated on presses, drills, or other equipment for use as the outer or inner shells of refrigerators and freezers. In addition, skilled Seeger machinists designed and built the tools and dies that were required for use with this machinery to do specific jobs in refrigerator production. Consequently the Metal Shop was the largest department in the plant, both in the floor space it occupied and the number of men it employed. The Metal Shop had approximately 50,000 square feet or about twenty-five percent of the total plant operating floor space. During normal production a total of about 500 men were employed in this department, working over three shifts of a twenty-four hour period.

The first impression a visitor had on entering the department was the great volume of noise that its many machines produced. After walking through the work area, the noise, however, soon lost out in importance to the impressive size and variety of machines that were in use. Here most of the machine operators, helpers, and truckers, working on three shifts, produced the 2,000 or more parts needed to keep the assembly lines moving. Here were all types of sheet metal machines such as the huge draw presses, the amazing Yoder Roll, shears, gang punches, power and hand brakes, punch presses, trim presses and tapping machines.

The procedures of the Shop required multiple operations to produce a finished metal part. Among the many operations undertaken in the department were, shearing, punching, notching, drawing, blanking, forming, gas welding, seam and projection welding, electric arc welding,

Voices *from the* **Plant Floor** *at* **Seeger**

No Coffee Breaks

"Seeger had a reputation as a frugal place to work. In 1934 there was a long line for job interviews. A secretary came with a handful of applications, laid them on a desk, and I took one and got back in line. I was asked at the interview if I could read a rule and blueprints. I was hired right away to work in the Metal Shop as a helper; my time clock number was 828.

"I worked the punch press machine. It had a bed eighteen inches by six feet and notched, pressed, and folded the outside of refrigerator casings. The factory was very busy. We worked forty hours a week and sometimes we were called to work overtime. My beginning wage was forty-two cents an hour.

"I worked at Seeger from 1934 until my retirement in 1965. Three hundred people attended my retirement party! Before World War II there were no coffee breaks during the day and we ate our lunch on the job. After the war, we had coffee breaks and the line closed down for lunch. I was thankful to have a steady job. After the war, my time clock number was 229 and at retirement 43."

Arthur Gehring

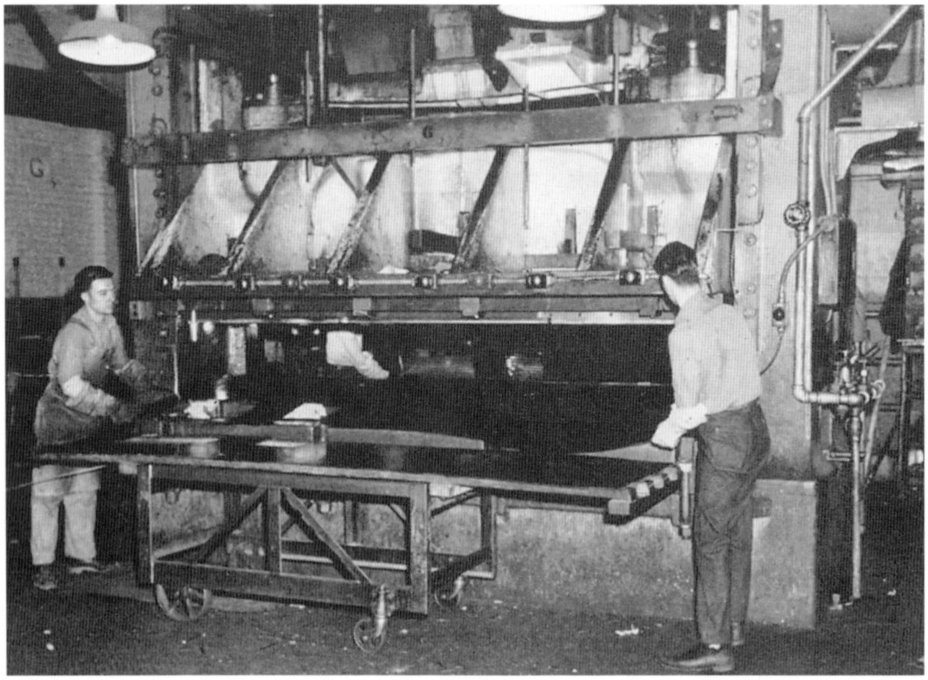

These machinists positioned a freezer side on the No. 1 draw press. After the freezer side passed through the press, two more helpers at the back of the press removed it. From a Seeger News *in the author's collection.*

soldering, and metal finishing. It was organized into several subdepartments in which a foreman was in direct charge. The subdepartments were known as Shearing and Punching, the Press Room, Commercial Assembly, Household Assembly, and the Welding Room.

In the Press Room the din was terrific! The men could hardly make themselves heard over the noise of the huge machines in operation. The visitor could see in all directions workers feeding lines of giant flat pieces of steel into presses. Hands reached for safety buttons that put the machines in operation. A dull "clump" then announced the meeting of specialized dies with the unshaped steel. Seconds later the operators removed smoothly

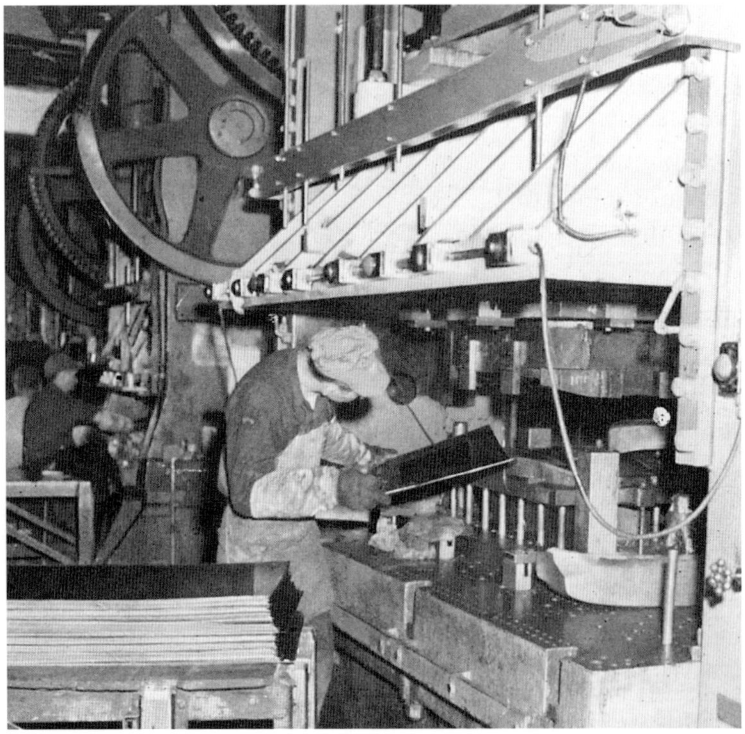

This was the No. 5 draw press. Here the machinist places a part in the press. A drawing compound was brushed on the outside edges of the sheet to prevent the steel from splitting when it was placed on the press bed and subjected to very high air pressure. From a Seeger News *in the author's collection.*

curved cabinet parts while a new piece of steel was guided into place. Thus, the making of a refrigerator cabinet began in the Pressroom on the Bliss, Minister, and Clearing Draw presses. The Bliss press, as a typical example, stood seventeen feet high and weighed forty tons. It could be operated at the rapid rate of fourteen strokes a minute at its highest speed. It bent and grooved sheets of 18-gauge steel with apparent ease.

With smaller dies and sheets of steel, three or four operations could be performed on each stroke of the machine. These simple facts provide an idea of how large, powerful, and versatile the draw press was at the Seeger factory. To maintain mastery over these giant machines, the operators had to be agile and sure in their movements at every moment.

Watching these men and their machines at work, the visitor sees only the flat piece of sheet steel go in and the curved side or top of a refrigerator came out of the draw press. But there is more to the real story than just this result. First the cold rolled sheet steel had to be sheared to size in order

Voices *from the* **Plant Floor** *at* **Seeger**

"The Best Job He Ever Had"

"Steve Mowchan was an employee, first of Seeger Refrigerator Company, and then Whirlpool Corporation for about 30-plus years. He and his wife came to St. Paul from the Ukraine after World War II. I guess it was about 1950 when some of the Displaced Persons [from Europe] were hired on the manufacturing and assembly lines. I was working in the Metal Shop, on a punch press when I first met Steve. He was working on the shears in the same department.

"One morning at break time, I went over to his shear to talk with him. He seemed to be mad about something at the time. I was interested to know where he had come from as I was thinking of researching our family tree, and his knowledge of the area in that part of the world would be of benefit to me. We never did get to talk about his part of the world because he was hot about the people he had to work with on the shears. I guess someone had called him a DP, and he felt it was a slur on him personally; which it was. 'They call me a DP, and that is what I am. I am DP. But, I am just as good as they are, and I am glad to get this job, and I will do the best that I can. Where I come from, I had a good job. It was a responsible job. I worked for the railroad and was a stationmaster. I had great responsibility on this job, and I liked what I was doing. Then the war came, and we ended up in Displaced Persons camps. (I believe in Germany.) These people here think because I am here that they are better than me. I had a better job over there than they have here. They are not better than me, I am not better than them.'

"There was a lot of this at the time, as the local workers had the feeling that the DPs would end up taking their jobs. It has always been that way, for the Swedes and Norwegians, the Poles, the Italians, the Finns, the Negroes, have faced this discrimination after landing on these shores, and it was not confined to workers in manufacturing plants. After a short time he made friends, and was a friend to many at the plant, and in different departments. He and his wife went to our church, and later found a church of their own. Their oldest son became a minister and Steve's children all went to college and found good jobs.

"At Steve's funeral, I talked to his son. I asked him about the display of Whirlpool pictures he had arranged on a table. There was his picture from the *Twelve O'Clock High* factory publication, his last official picture as he retired from the plant. Several other artifacts on the table were related to his job. His son said that his father was so proud of the fact that he had the opportunity to buy his house (of forty years) and send his kids to college. I believe the son told me that his dad had lived a great life and that his father told him that working for Seeger-Whirlpool was the best job he ever had."

Howard Pittlekow

to eliminate waste and to improve its maneuverability. Two workers then put it through a roller lever to clean it and make it more flexible. Then just before each piece was fed into the press, one man would brush the outside edges with a drawing compound that prevented the steel from splitting when it was placed on the bed under extremely high air pressure.

Under this bed—which was 60 inches wide, 108 inches long, and 8 inches deep—was a 265-ton air cushion that was built into the floor beneath the machine. The air pressure alone bent the steel. There was no heat involved in the process, just pressure, and lots of it. The air pressure that had been built into the floor under the press held the unformed steel up and away from the machinery so that when the top part of the press moved down, the steel sheet was free to bend or form into the desired shape that had been designed into the die or plate in the moving top part of the press. The plate or die could be changed in order to make different parts, tops and sides, liner bottoms, and doors.

After leaving the press, parts such as the shaped sides were taken to the next machine, operated by two men who would trim off the waste material along the edges. The only waste was the result of the drawing process that pulled the material from the sides up into the center and left excess at the four corners. It was then passed along to others who formed the edges, that is, rounded off and smoothed the raw edges. Three men at the next machine would trim the top of the end.

On the other side of the Pressroom were seven smaller presses. These machines performed various operations, but for our purposes we will consider the process through which a refrigerator door traveled. It presents an excellent picture of the sequence in which these presses operated. The first press turns and pierces the rear tiles, the flanges or turned portion that gave depth to the piece; the second trimmed and pierced the front stiles; the third pierced the bottom; the fourth formed the bottom corner; the fifth, formed the front sides; the sixth pierced the bottom rear stiles; the seventh formed the return flange on the front sides. Then, before the pieces went upstairs for metal assembly, the front stiles were taken to a power brake for squaring. The general flow of production called for pieces to be first cut to size, then progressed through the various draw and trim presses, then to the Metal Assembly where they were either assembled by welding or were metal finished. After inspection, they were placed on a conveyor that carried them to the Paint Finishing Department.

Among the veterans of the Metal Shop were several managers whose early days at Seeger dated back to the early 1920s, when the entire factory was located in the same main building in which the metal shop in the early 1940s covered two-thirds of the floor space. In 1942 Bill Tobiason, who had been with the company since September 26, 1921, was the most senior. Bill started on a job of nailing galvanized Ice Bunkers in place in iceboxes. He then worked continually in the Metal Shop. Bill was in charge of the special department in which replacement parts were made and special parts were fabricated. Second in line of seniority was Hank Reichert, who joined the company on March 21, 1924. He had experience working on Ice Bunkers and various jobs in the Metal Shop. The third man in line was Sam Lowrie, who began working with Seeger on December 17, 1924. Sam had the distinction of being the first press operator in the plant. When the Old Niagara Press No. 612-E was purchased about 1925, Sam was placed in charge of it. Later he held the distinction of having served as the first operator on practically every press installed in the plant. The fourth member of the management team was Joe Koening, who started with the company on September 1, 1925, by nailing Ice Bunkers in place as well as other jobs. Russell Magne was the fifth member of the management staff and he began working at Seeger on November 19, 1925. Magne's first job was the nailing of wood parts for the refrigerators. Later he became foreman of that department. For several years he worked in the Porcelain Department before transferring to the Metal Shop. The last of the department managers was Joe Pokorny, a welder, who joined the company on November 26, 1925.

Something out of the ordinary in the way of new equipment was the big draw press that was installed in 1941 and put in operation in the press department of the Metal Shop. This was the granddaddy of Seeger's battery of draw presses, towering head and shoulders over the other five that were at one time considered pretty good-sized machines. It measured eighteen feet from floor level to top, in addition to certain of its parts contained in a pit below floor level. An automobile might be parked on its bed or working area, which measured six feet wide and eleven feet long. The plate that made up the bed was ten inches thick and weighed 20,000 pounds. A forty-horsepower motor drove the press and developed 400 tons pressure. The total weight of the press was 100 tons.

Tools for Production

By 1928 Seeger machinists produced tools and dies to manufacture the parts fabricated in the Metal Shop.[3] The term "tool" was used in a general sense that both included dies and also referred to those parts of dies such as punches that were commonly called tools. Dies were the hard metal patterns used for forming or cutting out parts essential for the manufacture of refrigerators and later freezers. The dies varied in size from those that formed the small clips used to hold wiring in place in an appliance to those weighing

These men adjust a die for stamping the metal plate that can be seen in the background on the left. From a Seeger News *in the author's collection.*

several tons and able to turn out the rough shape of a freezer door.

The Tool and Die Shop designed and maintained all tools and dies for efficient production. Over the years the Tool Shop was primarily concerned with designing and methods while the Die Shop focused on maintenance and die tryouts. When an appliance model had been approved for production, the blueprints of each part were prepared. A copy of the print of each part to be fabricated at Seeger was sent to the Tool Shop. It was at this point that the tooling program was undertaken. The steps necessary to carry out the tooling for a new or redesigned product included the following: Methods of production, machines on which the parts would be fabricated, the sequence of operations, and the number of operations. While this was done, tools and dies were mentally designed. Then it had to be determined what standard tools could be used. In some cases used tools and dies may have been available that could be redesigned and then reused. New tools and dies were purchased from specialized sources when the preliminary planning indicated that the standard Seeger methods for tools and dies were not sufficient.

Voices *from the* **Plant Floor** *at* **Seeger**

"The St. Paul Division Will Do It"

"My mother and father were ballroom dancers and one of their fellow dancers, Peter Sands, was supervisor of the Tool and Die Department at Seeger. I was hired in 1936 right out of high school to work in Tool and Die. My starting wage was forty-seven cents an hour. Neil Griebenow was the Plant Manager and Kern Sosey the Plant Supervisor. Bob Ahrens, the Chief Engineer, was my mentor. Walter and Gus Seeger walked through the factory every day and said good morning. I became foreman in the Tool Shop and received a $100 a month raise in salary. Later I became the Supervisor of the Tool and Die Department. Between 1936 and 1939 or 1940 the factory employed more than 1,100 persons and worked three shifts. After the merger of the three companies in 1955, it was said that 'If you think it can't be done, the St. Paul Division will do it.'"

Floyd Wefring

Parts drawings that were received marked, "Released for Production" were then sent out for quotations. Quotations were normally requested from at least three sources. When the quotes were received, they were sorted for quality of tooling, fair price, and best delivery. Small tools and dies were usually placed in the Twin Cities area while large tools and dies were generally made in the East. The firms there had the larger equipment necessary to produce the large dies.

When the source had been established and the purchase order in the hands of the supplier, a design drawing of the tools or dies was requested from the supplier. The second step was a more thorough check of the design drawing to make certain that the tools would produce the desired part and on Seeger equipment. In some cases the parts themselves could not be efficiently and economically made on any tool. If that was the case, a request that the part be changed was sent to the Engineering Department.

If there was still some doubt as to the design, a third and final check was undertaken. After the design drawings were approved, they were returned to the tool sources and building proceeded. A further examination determined that the tools and dies were made to print, maintained the quality required, were delivered as scheduled, and made to suit the latest Engineering changes.

At this point that the Die Shop section of the department performed one of its most important functions. When the supplier delivered the dies, they had to be tested and proven for production. A setup of the die to be tested on the machines on which it was to be used was then made. Sample parts using the specified steel or other material were run off on the machine and carefully checked to make certain that they were according to print. If the parts were not right, adjustments were made on the die until the desired results were achieved.

After the dies were given a trial and the results found satisfactory, the Metal Shop was given an opportunity to prove their workability. The parts produced by the Metal Shop were run under production conditions and carefully checked. When production tooling problems arose, because the machine did not function properly, the material was too thick or too thin, the wrong type of material was being used, corners were breaking, tearing, bad scratches occurred, or holes did not line up, the Die Shop was called upon. They had to then provide the solutions to the problems. When the Metal Shop found the parts satisfactory, they were passed on to the Inspection Department for final approval.

Voices *from the* **Plant Floor** *at* **Seeger**

Working in the Tool Crib

"I lived on the East Side [of St. Paul], at 738 Burr Street, and saw an ad in the paper for a job at Seeger. For two years, 1947 to 1949, I worked in the Tool Crib, issuing tools to workers in various departments of the factory. We had to scrutinize the condition of the tools—good, fair, or poor. We also issued small hardware, such as screws, bolts, and nails and had to maintain a balance on the inventory. At the time Seeger manufactured refrigerators for many companies including Amana, Westinghouse, Frigidare, General Electric, Norge, and Kelvinator. Only the design of the doors was different and the nameplate."

Gerald Nielson

Die tryout was only a part of the work of this section, however. A great deal of time was spent in the maintenance of the dies and tools when production was underway. The men of this section not only checked over and repaired dies between production runs, but also made emergency repairs while the dies were on the presses. In addition they were responsible for making sure that the Metal Shop received the correct die from the hundreds in stock.

The Die Shop was set up on a three-shift basis and staffed with longtime, experienced employees. These personnel were accustomed to dealing

This photo shows the small die racks that held the many thousands of dies that the Seeger plant had in inventory. From a Seeger News *in the author's collection.*

with the many engineering changes that occurred during production runs and were prepared to make any necessary revisions in both old and new tools and dies. All men of the Tool and Die Shop had to be able to read parts prints, understand the tools and dies, and be able to make change-overs. They pointed with great pride to the fact that nearly all men in the department got their start in the production departments. Their skill was a result of their own efforts and a training opportunity that they had wisely accepted. Their contribution of Tools for Production was certainly an important one.

The Tank Room

The men of the Tank Room, with the aid of some of the most complex machines in the Seeger plant, formed one of the basic parts of a refrigerator and later the freezer: the liner.[4] With a gleaming porcelain interior surface, the liners served as the storage compartments of these refrigerated appliances. A wide variety of machines and operations were necessary to produce each and all models. It was a process that required a high degree of skill and ability from employees.

The unfolded liners were loaded on flatbed trucks and moved to the hydraulic folder. Here the sheet was formed to back and sides. A platform conveyor carried the partially completed liners to the Federal six-wheel seam welder. The operator stood at the left and placed a liner over the car-

In the Tank Room a platform conveyor carried partially completed refrigerator liners to the Federal six-wheel seam welder. The machine automatically completed all welds on the liner at the same time. From a Seeger News *in the author's collection.*

riage while the helper removed the completed one and placed it on the corner flanger. The bottom and top ends were inserted and clamped in place. The carriage rolled into the machine when the operator set it in motion. As the carriage rolled in place, the horizontal wheels descended and welded the top and bottom back. When these wheels reached the center of the weld, the left and right heads started to weld at the top and work down sides to the ends. All welds were finished at the same time and the carriage automatically rolled out. The helper on the seam welder removed the liner and put it on the corner flanger. The liners then rolled down a conveyor to the No. 1 and No. 2 trim presses and the side flanges were turned backward.

Liners were then placed on a platform where they were picked up by gas welders and placed in a welding jig. These men welded the four bottom corners to repair and strengthen them. From here the liners passed into the grinding booth on a roller conveyor where the job of smoothing the corners took place with an air-operated disc to remove burrs and imperfections left by previous operations. On the same conveyor the liners underwent another gas welding operation and the bottom back corners were strengthened and repaired before continuing on to the inspection station. Here the inspectors took over and checked the completed liners for open seams, correct dimensions, flaws in the metal, dents, rust, and any other imperfections. If necessary, flawed or imperfect liners then went back for repair. After this procedure, the completed liners that passed inspection were sent to the Porcelain Shop.

Using a portable spot-welder, an employee attached corner reenforcers to the right-hand end of the uppers and the left-hand end of the lowers. He then placed the completed work on the roller conveyor to the gas welding station. Two gas welders each welded one top left-hand corner and one top right-hand corner of the lowers. On the upper they each welded one top right-hand corner and completed the partially done left-hand corners.

To attach corner reenforcers to a freezer liner, this operator used a portable spot-welder. The parts moved to and from his workstation on a roller conveyor. From a Seeger News *in the author's collection.*

The Cleveland horn press was the next stop. This machine put the embossing or dents in the side corners to hold in place the coil tubes. Both uppers and lowers were embossed on this machine. At this point in the manufacturing process, inspection took over again to check for defects, open seams, incorrect dimensions, surface flaws, any dents, or rust in the refrigerator liners. Such imperfections were marked for correction. Few liners reached this stage that could not be used. The liners were then turned over to the Porcelain Shop for bunkering or loading.

6

Porcelain Enamel Finishing and Stocked Parts

THE EXTERIOR FINISH OF a Seeger product, from the earliest oak iceboxes to the latest *Coldspot* refrigerators, was a detail that reflected highly developed and refined technical processes. It was an application that served two purposes: visual attractiveness for the consumer's eye and long-term durability of the appliance. The porcelain enamel finishing process and Dulux paint applications were unique hallmarks of Seeger refrigerators.[1]

The Seeger porcelain enameling plant was one of the early enameling plants in the refrigeration field. In 1919, when it was established, the principal equipment for enameling consisted of a single, intermittent, box-type furnace, two spray booths, one drier, and a few small wooden tanks for pickling. At that time the enameling department only had to supply sufficient quantities of enameled parts to a crew of six or eight men in the refrigerator assembly department.

About a year later another intermittent furnace was installed for firing ground coat. The principal production of the enameling plant was refrigerator linings for the "iceboxes" then sold under the Seeger name. During 1924 the company constructed a new building and the plant layout changed to allow for rearranging the pickle room and the installation of three new muffle-type furnaces. In another five years, the demand for porcelain-enameled parts had greatly increased and an electric continuous furnace was installed. Now the plant was outfitted with the continuous electric and three oil-fired, box-type furnaces to handle the increased production that resulted from growing demand.

Porcelain enamel is a glass that is very closely related to ordinary glass and china. The difference is that extra ingredients are added to make it adhere to steel and to make it opaque with thin application. Instead of forming the porcelain enamel in molds as in glass manufacture, porcelain enamel is applied to a steel base in very thin coatings. This makes possible the manufacture of large shapes with a glass finish while at the same time reducing weight, cost, and danger of breakage.

The art of enameling on metal has been known since the time of the building of the Egyptian pyramids. Until approximately 1850 it was used, however, only for jewelry and other ornamentation, with copper and gold used as the base metal. Enameling on steel first began in Germany and was used as a finish for kitchen utensils. Since about 1914, its use as an interior and exterior finish on refrigerators became more and more common.

The characteristic that distinguished porcelain enamel from other finishes was that all of the raw materials were mineral in character. The primary ingredients in the enamel used on Seeger appliances were flint, feldspar, soda ash, potash, cryolite, fluorspar, and various metallic oxides. These minerals were ground to a fine powder, carefully weighed and mixed, placed in a furnace, heated (smelted) to about 2,000° F, at which temperature they became a molten liquid. This liquid was poured into cold water where it shattered into fine glass particles called "frit." Seeger once manufactured frit for its own use, but later it was purchased from outside suppliers.

Frit, clay, water, and small quantities of other ingredients were placed in a large cylindrical mill partially filled with china balls and again ground. The result was enamel "slip," which had the consistency of thick cream. The enamel slip was now ready to be applied, either by dipping the part to be finished,

This was one of several mills in the Porcelain Shop that was used to mix the ingredients for enamel slip. From a Seeger News *in the author's collection.*

or by means of a spray gun. Three kinds of enamel slip were used at Seeger; one colored dark blue or black, known as ground coat slip, and two other slip mixtures, both of which were colored white. One of these white slips was known as cover coat slip; the other was known as acid-resisting slip. The ingredients of the three enamel slips varied somewhat, as the ground coat slip adhered to the steel, the cover coat slip provided the color and finish, and the acid-resisting slip supplied resistance to attack by fruit acids.

The steel to which porcelain enamel was applied was milder, purer, and softer than ordinary cold rolled steel. This gave the advantage of better forming properties, and higher strength and stability to the heat applied in porcelain enamel processing.

After forming a part that was to be finished with enamel to the desired shape, the part was cleaned free from all dirt and grease in a boiling cleaning solution similar to dishwater but much stronger. The part was then washed, placed in a sulfuric acid bath to remove all rust, rinsed again, treated in a solution of nickel salts, rinsed in a mildly alkaline solution to remove all traces of acid, and then dried. This process, called "pickling" left the part chemically clean and ready for processing.

The next step in the application process required that the part be dipped in ground coat slip. The covered piece was then dried and additional slip was added, or "patched," where necessary by means of a spray gun. The part in process was then "fired" for about three and one-half minutes at a temperature of 1,550° F in a continuous conveyor furnace or an intermittent box-type furnace. The finely divided frit and clay in the slip coating were again melted while the part was in the furnace, which resulted in a fused sheet of glass forming over the part or ware.

After inspection and removal of blemishes, the cover coat was applied with spray guns. Where resistance to fruit acids was desired, for example, on the bottom of the inside of a refrigerator or a hydrator pan, a thin coat of acid-resisting slip was then applied over the cover coat slip. The part was again dried. The appearance after drying was similar to white cake frosting.

For assembly and ornamentation reasons, part of the cover coating was removed before firing by means of brushes. This operation, called brushing or stripping, was performed before the piece was again "fired."

After firing and inspection, the part was ready to be assembled in the refrigerator. In case the part was rejected for unsatisfactory appearance, an additional coating of cover coat slip was applied, or if this was not practical, the piece was prepared for complete refinishing by removing the porcelain enamel by means of a stream of sand or steel grit and air. This latter process was known as sandblasting.

Inasmuch as porcelain enamel is a glass, care was required in handling

Here an operator lowers multiple refrigerator liners in a bath for pickling, which leaves the parts chemically clean and ready for processing. From a Seeger News *in the author's collection.*

In this step a liner was lowered into a tank full of slip. Slip has the consistency of thick cream and was applied either by dipping or spraying. From a Seeger News *in the author's collection.*

After inspection and removal of any blemishes in the porcelain, a cover coat was applied with a spray gun. After the cover coat dried, it resembled white cake frosting. From a Seeger News *in the author's collection.*

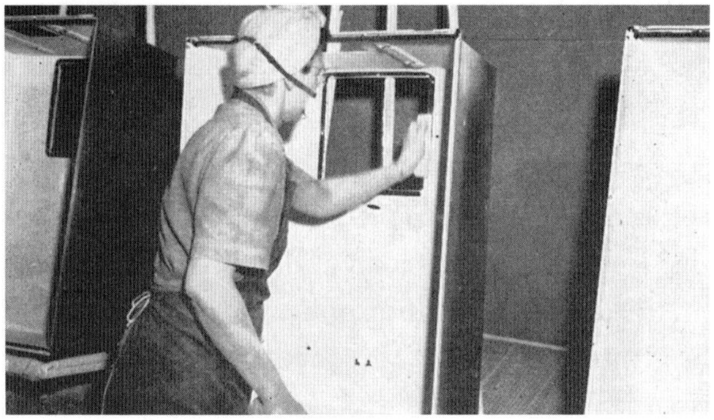

Before the liner could be fired in the furnace, part of the cover coat had to be removed with a wire brush. This step, which was known as stripping, made it easier to add ornamentation to the part and to complete assembly. From a Seeger News *in the author's collection.*

finished ware until the part was assembled. Only then did it gain its full strength and resistance to chipping, abrasion, heat, and cold. After assembling the component parts into a finished product, porcelain enamel, with reasonable care, was truly a "lifetime finish."

The next step was to assemble the component parts into a finished product. A long conveyor on which the employees "slaved" carried a series of frame works called cradles, which in turn carried the liners. The first step was to strip a liner of any excess porcelain for assembly and ornamental purposes. This process left all the holes in the liner clear of porcelain. If the porcelain had been left on, then when the screws were inserted during the assembly process and tightened sufficiently, they would have cracked the edges of the enamel finish.

The next step was to give the same attention to the top and bottom corners, a task accomplished with a brush and scrapper. Real neatness was required on this operation because these corners were in plain sight when the refrigerator door was opened.

This close-up shows the handwork that was part of the stripping step for refrigerator liners. From a Seeger News *in the author's collection.*

The next operation was stripping down the side of the liner with a five-row steel-bristle brush and reaming four holes on the side of the liner with a pencil-sized steel brush. The job of cleaning an indentation in the liner side was the final step in this part of the assembly.

The Dulux Finish for Refrigerators and Freezers

In 1935 Seeger engineers introduced an innovative process for providing a finish to refrigerators with the 1935 Sears *Coldspot* appliance.[2] This very successful new coating was the Dupont "Dulux" finish that was applied to some Seeger refrigerators. It was a very important operation in the sequence of steps involved in the manufacture of these cabinets. The purpose of this finishing operation was twofold: to provide the appliance with protection and superior visual attractiveness. The Dulux finish had to afford protection because Seeger used cold rolled steel in the fabrication of its refrigerator cabinets. This type of steel will quickly rust under the conditions of service for the average refrigerator. The Dulux enamel prevented this type of rusting. Given the demands of the consumer market, the finish of the refrigerator also had to incorporate qualities of beauty and attractiveness to enhance the

It the circulation room various paints were mixed and then sent under pressure to the spray booths where they were applied to the parts. From a Seeger News *in the author's collection.*

In this step the ware passed through a stage that cleaned, applied Bonderite, and then washed it before the primer and enamel coats were applied. From a Seeger News *in the author's collection.*

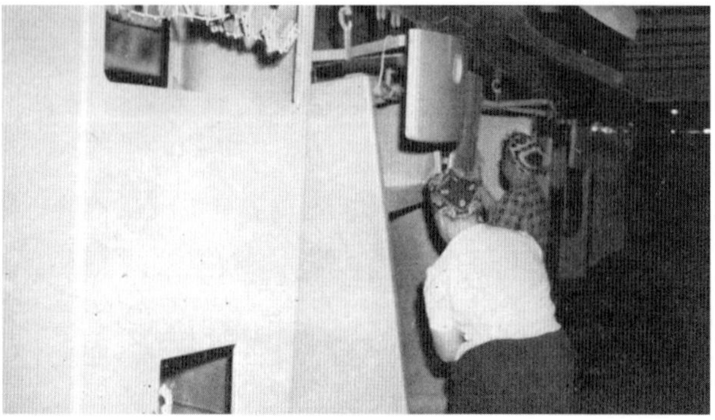

Here a worker removes particles of lint and dirt from the ware so that it is as clean as possible before priming and finishing. From a Seeger News *in the author's collection.*

Spray guns were used to apply the primer and finish coats to the ware. From a Seeger News *in the author's collection.*

appliance's sales appeal. Seeger managers readily understood that an appliance that was visually appealing would help stimulate buying and therefore encourage greater production. Thus they worked with the engineers to design a finishing system that would give Seeger refrigerators the necessary resistance to rust and a look that caught the eye of the consumer.

The Dulux finish applied to Seeger refrigerator cabinets had been carefully formulated to accomplish these objectives. It grew out of the Seeger Company's years of experience with the basic requirements of the refrigeration industry. This experience had shown that certain fundamental objectives had to be met. These general requirements for a refrigerator finish included the following:

1. Resistance to the moisture that condenses on the surface of the cabinet, which would otherwise cause rusting and blistering.
2. Adequate flexibility and adhesion to prevent chipping and cracking of the enamel surface during normal usage.

3. Grease resistance to prevent softening and erosion of the finish, particularly on the door, where it is handled.
4. Retention of color under the varying conditions of service.
5. Mar resistance to reduce marring and scratching.
6. Gloss and initial appearance to give eye appeal.
7. Ease of application and processing that enabled the finishing operations to proceed smoothly and with a minimum of production difficulties.

The Dulux finish on Seeger refrigerators combined the proper balance of all of these properties to produce a durable, attractive, and economical finish. Before the finish was applied, however, certain preliminary operations were necessary in order to prepare the metal properly for the finishing coats.

As they were received from the Metal Shop, the steel cabinets and parts that were to receive the Dulux finish were already covered with a light coating of oil and were further contaminated with drawing compound, shop dirt, ground metal particles, and similar loosely adhering material that was inherent in the metal-forming process. One of the most important principles in applying a finish to any type of object was to insure that the surface was entirely clean. Consequently, the first sequence of operations involved a thorough cleaning of the metal to remove all contamination.

Consequently the entire cleaning and subsequent finishing operations were conveyorized so that no handling of the parts was necessary after they were hung on the conveyor chain. First the ware passed through a chamber in which it was sprayed with a hot alkaline emulsion cleaner that was heated to a temperature of about 160° F. After a sufficient period of contact with this sprayed solution effectively cleaned all the contamination off, the work passed to a second spray stage where it was rinsed with hot water. Following this, the parts passed through a third rinse, again with hot water. At this point the metal was entirely clean and was also rinsed free of all traces of the cleaning solution. Failure to remove any traces of the cleaning solution would cause adhesion and blistering difficulties with the finish later on.

Experience had shown that for refrigerators that had to have a very high quality finish, this cleaning treatment alone was not adequate to give uniformly satisfactory results. Thus the parts that

Voices *from the* Plant Floor *at* Seeger

Hard Work in the Commercial Paint Shop

"For twenty years I worked in the Commercial Paint Shop at Seeger spraying paint on the display cases used by meat markets. There was also a paint shop for the domestic line of appliances. Later I worked in the Inspection Department reviewing parts as they arrived for use on the assembly line. The work was hard, but the people were good. We had 'hard time' parties at the East Side Commercial Club. My son received a college scholarship to study at Luther College in Decorah, Iowa. He is now a vice president of the Fairview Hospital. Three of my children worked at Seeger during their summer vacations, earning money to help pay for their college educations."

Darrel Borgstrum

were to receive the Dulux finish went through a further treatment of the metal to reduce the tendency toward blistering and corrosion. This was accomplished by passing the ware through a spray stage of Bonderite, under conditions specified by its manufacturer, the Parker Rust-Proof Company. This solution was operated at a temperature of between 110° and 130° F and actually changed the surface of the steel to a different chemical composition, which more effectively resisted rust and blistering. The next spray stage involved the rinsing of the Bonderized work in cold water. Following this the parts passed through a spray stage of hot dilute chromic acid in order to provide a still more uniformly resistant surface. At this point the steel was coated with a uniform dark gray crystalline coating, which served as an excellent base for the Dulux primer and enamel coats to follow.

The Bonderized parts were then dried by passing them through a short oven at a temperature of between 250° and 300° F. After these parts had cooled, they were then ready for tack-ragging and inspection. The tack-ragging operation was necessary to remove particles of lint and dirt that might otherwise have shown up in the finish coats and spoiled the beauty of the cabinet.

The treated parts were then ready for their first coat of finish, which consisted of a Hi Bake Dulux Primer. The purpose of the primer coat was to secure good adhesion to the metal and to provide a moisture and corrosion resistant base for the enamel coat. The vehicle or binder of this primer was a synthetic resin, manufactured from oil, glycerin, and phthalic anhydride, a chemical derived from coal tar. The primer was pigmented with materials that provided the highest degree of moisture and corrosion resistance. Using spray guns specifically designed for this purpose, the primer was applied to the metal parts. Prior to spraying, the primer had to be reduced or thinned to the proper consistency in the paint mixing room and then was pumped through a continuous circulating system that supplied the individual spray guns. Compressed air was also supplied to these guns. The mix of air and primer atomized or broke up the paint into small particles, which produced the smooth uniform appearance that was characteristic of the finishing coats. The primer was baked at a temperature of approximately 350° F for a specified time in order to harden or polymerize it. Because the primer had been formulated to hold a light yellowish-brown color at this high temperature, these parts could be easily distinguished.

After the primer coat was baked and the work had cooled down sufficiently, it was scuffed very lightly with fine sandpaper to remove dirt nibs and any other roughness. Then the surface was tack-ragged to remove all sanding dust and other dirt.

The parts were now ready for the finish or enamel coat. This was also applied by spray, in the same manner as the primer. The enamel was baked too, but at a lower temperature than the primer, the peak being approximately 300° F. This coat incorporated the qualities of attractiveness, whiteness of color, gloss, and resistance to grease, staining. and marring which were so important for a finish of this type.

The Dulux enamel used on Seeger cabinets was based on a synthetic resin vehicle, made from oil, glycerine, phthalic anhydride, and urea formaldehyde. The pigment used in this enamel formulation was one of the whitest known—titanium dioxide, which was obtained from black-colored sand called ilmenite. This enamel was remarkably white and retained its whiteness to a high degree during baking and also during service. Each batch was carefully tinted or shaded to match a definite color standard so that the production color would not vary over any period of time.

All spraying operations were carried out in special areas that were supplied with filtered air that was kept under a slight positive air pressure in order to prevent the entrance of dirt into the spray booths. A water curtain at the back of each spray booth captured the overspray, or that portion of the sprayed material that did not strike the ware. The accumulation of water and enamel that was collected by the water curtain, which was called sludge, was later reprocessed by the DuPont Company and could be reused so that very little material was wasted. The spray-gun application of the primer and enamel films was very carefully regulated in order to attain as uniform a film thickness as possible. This part of the quality control process involved the constant monitoring of the viscosity or tendency to flow of the material, as well as control of the pressure of the air in the spray guns, and the velocity of the coating materials that were supplied to the spray guns.

Likewise the ovens used for the baking of the primer and enamel coats had to be carefully regulated, since the amount of bake which each of these coats received had to be adjusted to give the best results. Special instruments automatically maintained this oven control after the proper setting had been established by means of accurate temperature tests.

A great many other things were involved in carefully controlling the finishing operations, such as the proper hanging of the parts from the conveyor, the necessity of never touching bare metal with ungloved hands after it had been cleaned, the maintenance of good housekeeping in order to reduce dirt that might otherwise get on the finish, the establishment of uniform spraying conditions, and other factors. All of these relatively minor points had to be taken into consideration, however, since attention to these details could mean the difference between turning out a satisfactory finish and one that failed to meet the company's usual quality measurements. The

DuPont Company sent service technicians to the Seeger factory at frequent intervals to assist in checking over all finishing operations and to promote the most efficient use of their products.

After the parts had been finished, they were inspected prior to assembly so that any that were defective with respect to uniformity and quality of finish could be rerouted for repair and the application of another enamel coat. Although great care was taken during the assembly operations to avoid the possibility of any mechanical damage to the finish, a screw driver might slip occasionally, or the finish might be gouged slightly for one reason or another, in which event the finish had to be patched, provided that the location and area of the damaged spots fell within certain standards. This patching was carried out with DuPont "Duco," using a white lacquer, which matched exactly the color of the baked Dulux film. This material air-dried in a short time, and after rubbing and polishing the patch, it was difficult to tell where such repairs had been made.

Underlying all these production operations was a constant program of testing. The DuPont Company carefully pretested all the Dulux finishes which were supplied in order to maintain uniformity of quality. This testing

Two pickers or issue clerks examined a rubber gasket to make sure that it conformed to a customer order before the gasket is packed for shipment. From a Seeger News *in the author's collection.*

program included the raw materials from which the finishes were made, the intermediate resins, pigment bases and solutions which entered in to the various manufacturing operations, and also the complete evaluation of the finished products for physical and working properties. These finishing materials were further tested in the Seeger Laboratories in order to insure that they would perform satisfactorily under the operating conditions in the Seeger plant and as a further check on the quality that Seeger had established and maintained throughout the years. Specifications prepared by the Engineering Department covered all of the manufacturing steps that have been described. These specifications established the proper procedures and the standards that were required. In addition the Seeger Engineering Laboratories continually checked finished production ware to insure that these standards were being met.

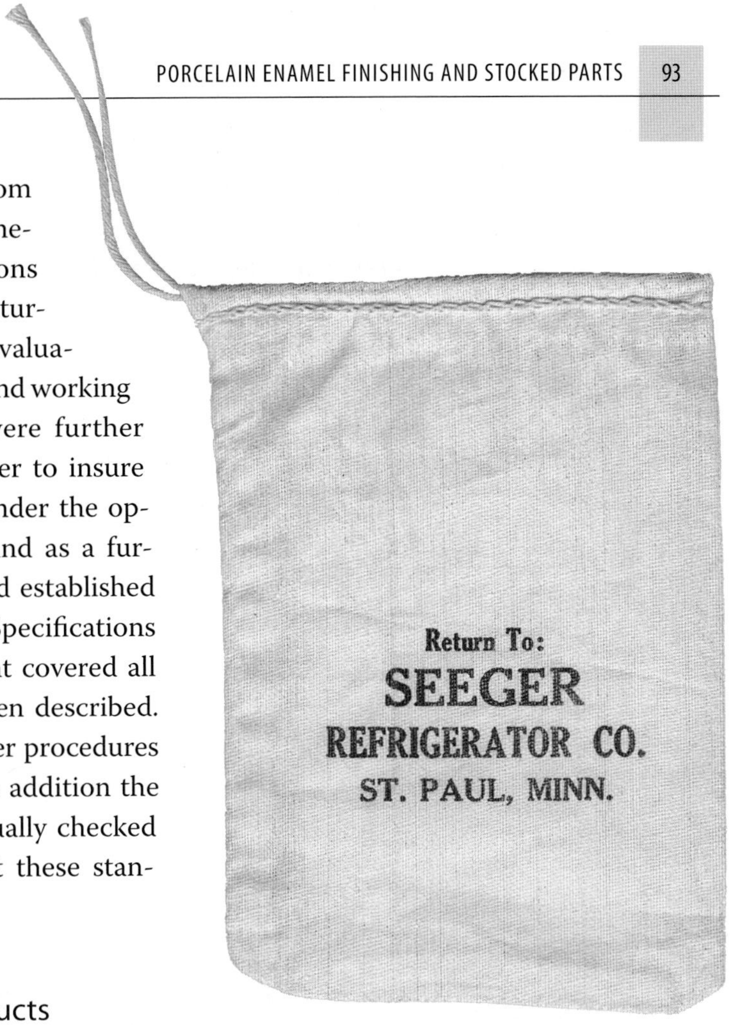

A lightweight linen bag which the Express Department used to ship small parts to customers. From the author's collection.

An Inventory of Parts for Seeger Products Manufactured between 1928 and 1955

As refrigerator production volumes increased over the years, the Seeger factory developed and maintained an extensive inventory of the parts for products that it manufactured from 1928 until the merger with Whirlpool and RCA in September 1955.[3] During that period, the staff in the Express Department was constantly involved in the collecting, storing, and shipping of all sorts of parts for more than twenty-five yearly models of refrigerators that Seeger manufactured under half a dozen trademarks of other appliance merchandisers. The department had on hand the service parts necessary to replace almost every working part that might wear out in all the styles and makes of refrigerators that the Seeger Company had manufactured after 1928. Those that were not on hand could be made with reconstructed dies or ordered from affiliated companies. The department maintained a complete file on blueprints and serial numbers of refrigerator parts and styles from 1928 to 1955. Their tasks were complicated because the Express Department filed and maintained at least 3,000 items of varying shapes and sizes. The inventory included materials for the many models of residential and commercial cabinets, both left- and right-hand door styles on all refrigerators, plus all the types of freezers that were produced by the company.

The stock of parts carried was determined by normal usage in the field during the manufacturing period. Replacements were then stocked after agreement with the manufacturer for whom the product was made, say, for Sears, Wards, Amana, or Frigidare. A three- or six-month stock was sufficient during manufacture. For discontinued models a one-year supply was held. If the dies were to be scrapped, then a life usage figure was determined and parts were run off on that basis. Often a member of the department had to check through a file of blueprints to determine what the customer needed because many parts that were ordered were for refrigerators manufactured during the previous twenty to twenty-five years.

7

1934–1941: The Impact of the Sears *Coldspot Super Six*

FROM THE FOUNDING OF Seeger in 1902, the line of products manufactured included icebox cabinets for residential and commercial use. The familiar oak icebox was a product that Seeger made for the home market until about 1926 while the company also manufactured special refrigeration cabinets for creameries, butcher shops, hotels, and cold storage facilities. Seeger also installed its patented "Siphon System" of refrigeration in railroad freight cars for several railroad companies including the Great Northern, the Northern Pacific, the Chicago, Milwaukee & St. Paul, the Chicago, Burlington & Quincy, the New York Central, and the Pennsylvania. In addition the company established the production norms for steel refrigerator cabinets when in 1916 it began to manufacture the product for General Electric. The Seeger cabinet rapidly became the model and standard for the emerging residential refrigeration industry. Throughout the 1920s the company consolidated and expanded its position in the field.

By 1930 the impact of an escalating economic depression in the United States spread from Wall Street to Arcade Street. During the course of county-wide economic collapse many industrial firms—both great and small, local, regional, and national—fell by the wayside. In St. Paul the Seeger Company's immediate task was to survive in the face of steadily adverse market conditions. At the time an electric refrigerator was considered a luxury item rather than a necessity. Thus the decreased purchasing power of consumers nationwide was one of the first signs of the worsening economic conditions in the United States. Despite what happened across the country, production and sale of the Sears *Coldspot* refrigerator increased year after year. Beginning in 1928 when the Seeger Company got its first order to manufacture the Sears *Coldspot* refrigerator in alliance with the Sunbeam Electric Manufacturing Company of Evansville, Indiana, production volume at Seeger went up dramatically. Thereafter, about 90 percent of the company's production was the *Coldspot* appliance.

Sears, Roebuck and Company's celebrated 1935 Coldspot Super Six. *Although the interior space of these refrigerators varied slightly depending on the specific model, all had an interior capacity that exceeded six cubic feet, which led to the use of* Six *in the model name. Whirlpool Corporation photo.*

This is the 1938 model of the Coldspot Super Six. *Here the design of the door handle has changed substantially from the 1935 model, the hinges for the door are no longer visible from the front, and the feet on which the appliance rests are fancier than they were in the older model. Whirlpool Corporation photo.*

During the first five years of the 1930s, Seeger managers changed the existing Arcade Street factory to bring about increased productivity, not by adding more manufacturing space, but by more efficient utilization of existing facilities. As production increased and changes in the plant took place to accommodate these increases, a similar change also took place in the cabinets themselves, a trend toward all steel construction, more flowing and graceful lines, and the incorporation of decorative trim and accessories to create a cabinet of beauty and art.[1]

In 1934 Sears, seeking to streamline the style of *Coldspot* refrigerators for an ever-expanding consumer population, retained the flamboyant designer Raymond Loewy to redesign the appliance. Loewy made the refrigerator into a rectangular component entirely usable for storage and gave detailed consideration to all hardware as a part of the total design. The introduction of the 1935 refrigerator, called *Coldspot Super Six,* greatly influenced the subsequent look of all kitchen appliances. The Sears Mail Order

Catalog of that year noted that the *Coldspot Super Six* was "A Refrigerator out of the ordinary in design . . . years ahead of any other refrigerator in beauty and style, the equal of any in mechanical perfection and with every modern convenience." It measured 58-⅛" high, 30" wide, and 26" deep and its actual storage space was 6.20 cubic feet. The refrigerator weighed 390 pounds and cost $134.50 in cash, the equivalent of $1,856.76 in 2005 dollars, or $5 down and $8 per month.

Loewy's efforts were brilliantly successful. His design won first prize at the Paris International Exposition of 1937. Later Loewy recalled in his autobiography that the Sears, Roebuck success with the *Coldspot Super Six* was the turning point in his career. It launched Loewy as one of the foremost industrial designers of the twentieth century.

In addition to all the accolades Loewy and the *Coldspot Super Six* subsequently received, in 1935, the first year it was in production, the redesigned refrigerator became a resounding commercial triumph for Sears and Seeger. Sales of the appliance in that year increased to 275,000 against 65,000 for the previous model. This sales success was more, however, than a result of

Raymond Loewy

Born in France in 1893, Raymond Loewy was one of the foremost industrial designers of the twentieth century. At the age of fifteen his design for a model aircraft brought him his first public award. After serving in the French Army during World War I, he migrated to the United States in 1919. Loewy first lived in New York City and employed his talents and skills as a window designer for department stores. By the late 1920s he received commissions for such firms as Westinghouse and the Hupp Motor Company, the latter for the styling the company's automobile, known as the Hupmobile.

A 1919 portrait of the industrial designer Raymond Loewy (1893–1986). Library of Congress photo.

In 1934 Sears, Roebuck and Company retained Loewy, who by this time had moved to Chicago, to redesign the *Coldspot* refrigerator with the assistance of the Seeger Refrigerator Company's Engineering Department. The 1935 *Coldspot Super Six* model was the culmination of the innovative design and technical project. This kitchen appliance represented fundamental precepts that drove Loewy's creative efforts. To Loewy, "Good design is not an applied veneer," and it should represent "beauty through function and simplification." The newly designed refrigerator was not only appealing to the critical eye of the consumer, but a resounding marketing success for Sears and a manufacturing bonanza for Seeger. In fact the *Coldspot* fulfilled another Loewy maxim that "the most beautiful curve is a rising sales graph." He remarked in his autobiography that the Coldspot project really launched his national and international career as an industrial designer.

The legacy of Loewy's creative projects is recalled in a wide range of popular consumer products, such as the logo for the Hoover Vacuum Company, the line of Frigidare kitchen appliances, the Pennsylvania Railroad locomotive, the bullet nose Studebaker automobile of the 1950s and 1960s, the logo for the Lucky Strike cigarette package, the glass Coca-Cola bottle, and the logos for the Shell and Exxon oil companies.

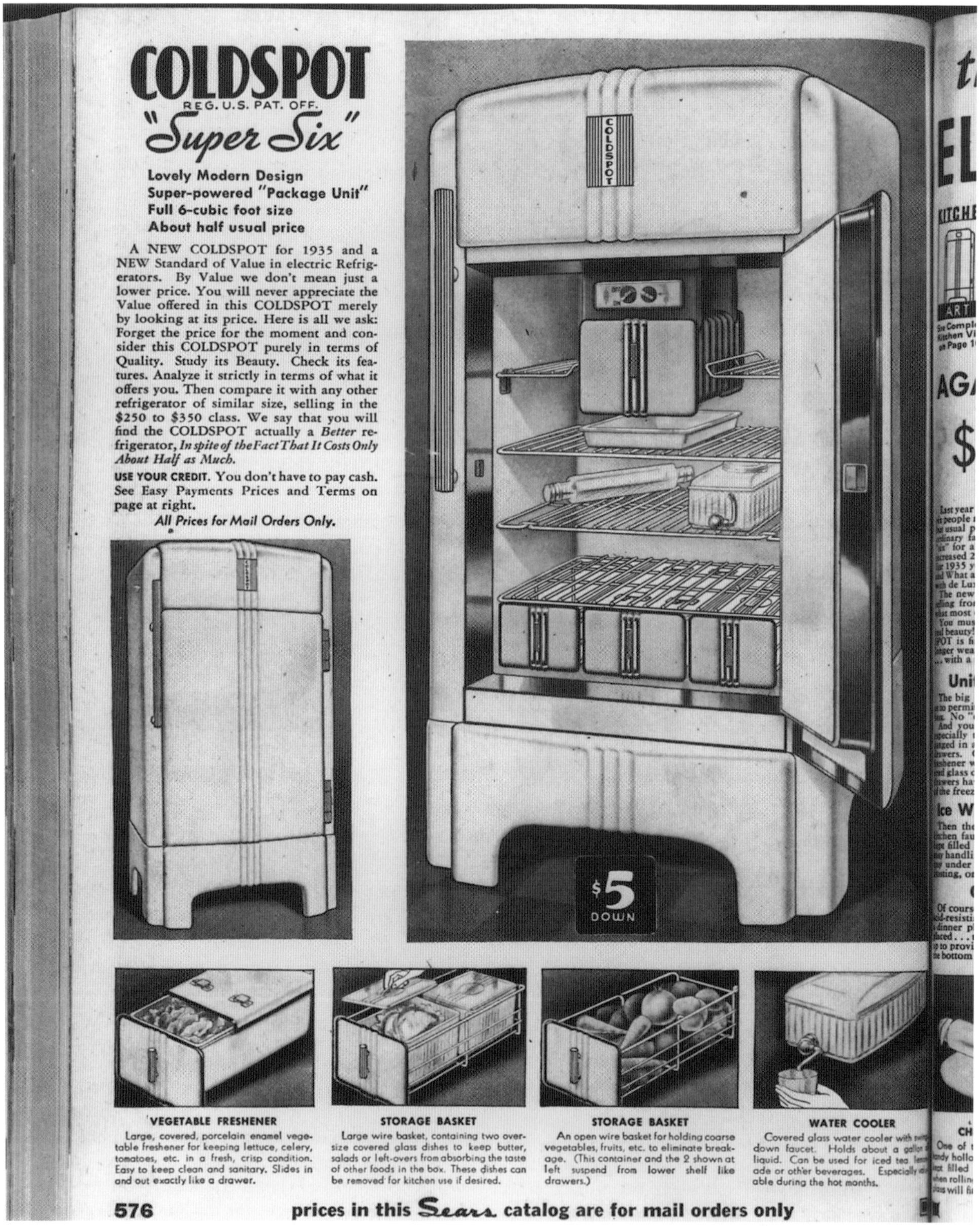

The Spring–Summer 1935 Sears catalog carried this advertisement for the Coldspot Super Six *refrigerator. From a Sears catalog in the author's collection.*

the Sensation of 1935 in
ELECTRIC REFRIGERATORS

Think of it! A refrigerator, out of the ordinary in design . . . years ahead of any other refrigerator in beauty and style, the equal of any in mechanical perfection and with every modern convenience . . . a refrigerator that will be the envy of your friends.

A world famous designer developed the new streamline cabinet to conform to modern home decorative ideas. Well known engineers designed and built the *"Package Unit"* that insures continuous, trouble-free service. Famous Home Economic advisors selected the equipment and designed the interior for the utmost efficiency.

Then . . . what a surprise as you read the low price and easy terms! We've purposely priced it to meet the purse of the average family, so that *no one can afford to be without* the 1935 Super-Six COLDSPOT . . . America's outstanding refrigerator value.

AGAIN, a "Six" at the Price of a "Four"

$134.50 CASH

Last year we found that the average family of from four to people required a refrigerator of six cubic foot capacity, usual prices for that size were out of the reach of the family. We did something about it—we sold a for about the usual price of a "four" and our sales increased 287%. We will follow the same policy this year. 1935 you again get a "six" for the price of a "four"— What a "six"!—newly designed, better than before and de Luxe equipment throughout.

The new 1935 Super-Six equals any refrigerator made from $250 to $350, and our price is about half most others ask.

You must really see the COLDSPOT to appreciate its beauty! And, to emphasize that beauty of design COLDSPOT is finished with glistening, snowwhite DuLux . . . wearing, glossier, easier to keep clean than lacquer . . . with a touch of chromium provided by the hardware.

Uniform Freezing Throughout

The big 105-cube freezing unit is right in the center so permit perfect circulation of cold air to all parts of the No "dead" corners or pockets.

And you'll really appreciate the many other features, especially the three bottom compartments which are arranged in a row and suspended from the lower shelf like drawers. One is a large porcelain enameled vegetable container with lid . . . one is fitted with two oversize, covered glass dishes . . . and one is an open wire basket. These drawers have two-tone streamline fronts to match the front of the freezing unit.

Ice Water Whenever You Want It

Then there is the water cooler . . . as handy as a third kitchen faucet . . . a hollow glass rolling pin that can be filled with cold water so as to firm your dough for handling and so that it won't stick . . . and a large glass under the freezing unit to catch the water when defrosting, or to hold surplus ice cubes.

Clean, Sanitary Interior

Of course, the entire interior is porcelain enamel, with acid-resisting porcelain bottom . . . as easy to keep clean as a dinner plate. Notice how conveniently the shelves are placed . . . the two alongside the freezing unit can be folded to provide space for tall bottles. The second shelf from the bottom is a split shelf which can be removed to provide room for tall bottles, roasts or melons. And . . . no fumbling in the dark, because the interior is brilliantly lighted the moment the door is opened. To keep the cold IN and the heat OUT the entire box is heavily insulated with *Dry-Zero,* the most effective insulation known.

"Keeps Milk Fresh 14 Days in Desert"

"We live in the heart of the Mojave desert where the temperature is often as high as 120° in the shade. Have kept milk fresh 14 days during summer," Chas. A. Bly, Jr., Needles, California.

Few COLDSPOTS are required to operate under conditions as severe as this. Yet, it is a source of immense satisfaction to realize that the famous *"Package Unit"* operated at peak efficiency under these trying conditions.

"Saved $100 Buying a Coldspot"

"Saved $100 on the COLDSPOT, so you can figure out for yourself that we are ahead considerable," Chas. S. Fleischmann, Oil City, Pennsylvania.

That saving on first cost is a considerable item, but when you also have the efficient *"Package Unit"* to insure years of satisfaction . . . that's real economy.

30-Day Trial (Cash or Easy Terms)

We'll gladly send you a COLDSPOT on 30 days' trial, to test and compare right in your own home. Send the cash price, or the deposit (using the Easy Payment Order Blank in back of the catalog). If it's everything we claim for it . . . send the small monthly payments until easy payment price is paid. If not, return it and we'll return your money.

MEASUREMENTS: 58$\frac{3}{16}$ in. high, 30 in. wide, 26 in. deep. Actual storage space 6.20 cubic feet. Shelf space 13.10 sq. ft. For 110-volt, 60-cycle alternating current. For any other cycles, or for direct current add $5.00 to prices shown. State type current you use, and if alternating, state if 25, 50 or 60 cycles. This information is on your meter. Shipped from Philadelphia or Boston. Shpg. wt., 390 lbs. To estimate freight charges see page 848.

22 HM 3303—New Super-Six COLDSPOT Refrigerator. Cash Price $134.50
Easy Payment Price, $5 down, $8 Monthly . . . $147.95

22 HM 3304—Exactly same as above, except full porcelain enamel finish, outside as well as inside. Shpg. wt., 450 lbs. Cash Price $154.50
Easy Payment Price, $10 down, $9 Monthly . $169.95

22 HM 3305—8 cu. ft. net, full porcelain enamel refrigerator, exactly the same as our 22 HM 3304 except larger. Measures 62¾ in. high, 33¾ in. wide, 26¼ in. deep; four big shelves; shelf area 17.83 sq. ft.; freezes 138 cubes in five freezing trays.
Cash Price (Shpg. wt., 495 lbs.) $174.50
Easy Payment Price, 10 down, $10.50 Monthly $191.95

Use Easy Payment Order Blank in Back of Catalog

"SUPER-POWERED" PACKAGE UNIT

Purposely built about ten percent oversize to take care of the extra ice cubes and to provide enough refrigeration, even in the hottest summer weather without effort. A super-feature for the 1935 COLDSPOT.

"FINGER RELEASE" ICE TRAYS

Just a flip of the finger to release any of the five trays. No pulling. No prying. Freezes a total of 105 cubes in record time. Five trays—one holds 42 cubes, another 21 and three hold 14 each.

FEATHER-TOUCH DOOR OPENER

This is the beautiful, chrome plated "Feather-Touch" door opening bar, located at just the right height for hand or elbow. It opens the door with just a touch so that with full hands you can open the door without effort.

CHILLED ROLLING PIN

One of the added conveniences is this handy hollow glass rolling pin that can be filled with cold water. You'll find, when rolling out your dough that the chilled glass will firm it and keep it from sticking.

SPLIT SHELF FOR BOTTLES

Here's something mighty handy . . . the split lower shelf. Take part of it out and you will have extra room for tall bottles, roasts, fowl or watermelons. Many a time you will appreciate this convenience.

SEMI-AUTOMATIC DEFROSTING

Turn the dial to "Defrost." The frost around the freezing unit will melt into the glass tray . . . but the ice cubes remain solid. When defrosting is completed the current turns "on" again automatically.

10 FREEZING SPEEDS

Very slow to very fast. Regulates easily to exactly the temperature needed, depending on the weather. Nine regular Freezing Speeds and one "Quick Freeze" for use when you want ice cubes in a hurry.

write Sears personal service for suggestions 577

superior industrial design and shrewd marketing overcoming the national hard times. To understand how this happened, we need to briefly examine the process used to manufacture a Seeger refrigerator.

Most electrical refrigerator cabinets regardless of the name or label on them, in outward appearance at least, were similar in construction. They consisted principally of an exterior case that was either self-supporting or attached to a wood frame, and an inside shell or liner, surrounded by suitable insulating material. The thickness of the metal used, the kind and quality of finishes, the kind and thickness of insulation, the moisture proofing method, the kind of wood frame, the style, arrangement, and added conveniences offered were the way in which most cabinets differed.

Over 6,000 inventory items were required to supply the company's production and maintenance departments. The bundles of steel delivered to the factory were broken in the storerooms, the sheets separated, and then loaded onto hand trucks. After the truck was loaded, the steel was moved to the shearing section. The steel was issued to the shearing section on the basis of a shearing schedule and shear operators cut the necessary pieces according to specifications. The pieces were then taken to the press section where they were formed into sides, tops, doors, and so forth. On occasion several operations were necessary to complete a part for assembly.

Smaller parts were made in punch presses. Excess steel sheared off the larger pieces during the forming operations was often utilized in making these smaller parts. Making these items might also involve multiple operations. When the completed parts were ready, they were moved by hand trucks to the Metal Assembly Department. In this department the sides and tops of a refrigerator cabinet were spot-welded together in a specially built machine, forming the main part of the refrigerator shell. The shell was then passed to another crew who added the back by a spot-welding process and additional parts were also spot welded to the refrigerator shell. Using some additional spot-welding and with the aid of certain jigs, some necessary holes were drilled in the shell. The shell was then nearly completed.

At this stage the metal finishers took over and ground off any rough spots that could be found on the shell. This was an especially important task and could be performed only by experienced workers. The initial metal finishing completed, the Inspection Department carefully checked the shells for any flaws in the metal and marked them for further work if necessary. Those found defective were set aside.

After the metal finishing was completed, the exterior parts were loaded on a conveyor leading to the Dulux Department. The parts were given a final check by inspectors. Any defective pieces were again set aside for appropriate action.

These two pages are from a 1938 Sears advertising booklet that explained all the advantages of having a Coldspot Super Six *refrigerator in the home. The ad copy on the left-hand page touts the refrigerator's economical use of power and the right-hand page promotes the capacity of the appliance. From a Sears advertising booklet in the author's collection.*

This aerial photograph shows the Seeger Refrigerator Company plant on Arcade Street about 1930. The railroad spur that served the factory is visible toward the bottom of the photo while residential housing on St. Paul's East Side dominates the area at the top of the photo. Photo from the author's collection.

The main interior part of the refrigerator was the liner. This was assembled in the gigantic automatic seam-welding machine after the parts were in place. The completed liners were inspected after the metal finishing had been done. All the liners that passed this quality control check were then loaded on a conveyor that led to the Porcelain Department.

The liners reached the Porcelain shop and were placed in racks. The racks were picked up and lowered into the pickling vats. The acid solution in the vats removed all oil, dirt, or rust in preparation for the ground coat. Following the application of the ground coat, a grayish paint-like substance, was applied to the liners. They were then dipped into a vat by hand and set onto racks allowing the excess to drain. The liners were then placed on a conveyor chain that passed through a continuous oven. The ground coat became the black enamel-like material found on the liners.

With the ground coat baked on, the liners passed along the chain to the spray booths where a coating of porcelain was applied to the interiors. From here it passed through a drying oven. The drying ovens caused the porcelain to become firm. As the liners left the ovens, the strippers removed certain portions with the use of brush and scrapers for ornamental and assembly purposes. The stripping operation completed, the liners passed into continuous ovens to bake on the porcelain. Porcelain was used rather than enamel because of its greater resistance to wear and damage from food acids. The completed liners emerged from the continuous ovens where they were inspected. The perfect liners were removed and placed on the conveyor for the journey to the assembly line.

Panelyte, a composition paper material which was used for molding and the inside panel of the doors, was another item that was important to the appearance of the refrigerators. Inside door panels were loaded onto racks after they had been inspected and sanded. The panels were cleaned to prevent any dirt or dust from sticking to the enamel before they entered the spray booth. The panels were removed from racks after the spraying and drying had been completed. These panels were inspected, loaded onto hand trucks, and then moved to the assembly line for use.

The exterior parts passed from Metal Assembly to the Dulux Department where they passed through a washing process during which all dirt and grease was removed and any excess moisture was removed with compressed air. The parts were then dried before reaching the prime coat spray booth. The first coat of enamel was applied and dried in continuous ovens. On emerging they were wiped and sanded. The parts then passed into the finish coat spray booth where they were carefully given a final coat of enamel. Each sprayer had a particular part of the shell to paint. Continuous drying ovens insured a lasting quality to the enameled parts. A chain then carried the freshly enameled parts from the ovens to the assembly floor where they were removed from the chain. A careful inspection was made before parts were allowed to pass to the hands of the assembly crews. A belt conveyor carried the shells to the first operation while other parts were carried to the necessary places by overhead conveyors.

The operations were rapidly completed once the shell was placed on the assembly conveyor. Tar was applied to the seams and a water test was used to insure a complete sealing of the refrigerator shell. Spun glass insulation was then placed in the shell in several steps. The last of these steps added not only insulation, but also an insulite partition to hold the liner in place. A liner from the Porcelain Shop was placed in the shell. Prior to this step, each liner had been fitted with the wire for the interior electric light connection and some of the shelves for the liner. These assembly-ready liners were then anchored in place and the molding that covered the gap between

This 1934 advertisement for Seeger's commercial cabinets and display cases appeared in a trade publication called The Progressive Grocer. *From a Seeger advertisement in the author's collection.*

the shell and the liner was attached to the both the inside and outside assembly. These shells and liners units were then installed in the refrigerator boxes, which already had a separate freezing compartment attached to them. The nearly assembled refrigerators were next uncrated on the floor below and delivered to the assembly line on a conveyor chain. Additional shelves and crisper pans were installed in the refrigerator. The movable parts were then padded and securely taped in place to prevent any damage from occurring while the refrigerator was in transit.

At this stage of the process, a crew then assembled all the door components. The empty door shell started its journey at the front of the assembly area and continued around to the left while the insulation, inside panel, hinges, and so forth were added. When the door was completed, it was removed from the line and hung on the refrigerator at a nearby point. The completed refrigerator box then passed through a series of rigid inspections before moving on to the craters. Any defective boxes were moved for correction of the faults. If the refrigerator passed the inspection, the craters moved into action. Every precaution was taken to see that the completed article was properly padded and covered so that no damage could occur during its transportation.

As early as 1911 Seeger began purchasing the hardware for its line of iceboxes from the Dent Hardware Company of Fullerton, Pennsylvania.[2] Richard Hagberg, who began his long career with Seeger in 1941, recalled the company's production supplies were provided by an array of firms. Inland Steel and U.S. Steel companies supplied the steel, paint came from the DuPont Company, Monsanto Chemical Corporation provided insulation for cabinets as well as foam and plastic pellets, the National Lock Company manufactured the lock assemblies and handles for refrigerators and freezers, Superior Plating supplied parts for dishwashers and vacuums, Tecumseh Corporation made the freezer condensers, Waldorf Paper Products and Atlas Plywood supplied the cartons for transporting the appliances, and the Conwed Corporation of Cloquet, Minnesota, provided the wooden pallets used on the factory floor and in the warehouse.

The exterior case of the *Coldspot* Dulux cabinet was die stamped from 21-gauge, deep-drawing, auto body steel, the finest steel obtainable for refrigerator cabinets. The formed case was then spot-welded, and the DuPont Dulux finish applied. Seeger was one of four companies in the United States licensed by DuPont to apply this new durable enamel. After this process was accomplished, the case was bolted to the refrigerator base, which was a separately stamped and welded unit. This completed unit or exterior case was constructed so that the compartment for the unit was self-supporting.

All *Coldspot* cabinets had one-piece, acid resisting, porcelain linings. All

Voices *from the* Plant Floor *at* Seeger

Working at Seeger Refrigerator Company in the Late 1930s

Ernst John Kulenkamp. Photo courtesy of the Kulenkamp family.

"Up until this time [circa 1936] now there wasn't all this modern stuff like we got now [circa 1987]. There was no electric refrigerators, or televisions or anything. Everybody had iceboxes. So the electric refrigerator was coming into its own. Sears & Roebuck, and Montgomery Ward, and all them were ordering all the electric refrigerators they could get to sell because it was just booming. Everybody wanted one. Seeger Refrigerator Company built the cabinets for these things and they were so busy. They were going day and night. Of course, I did not know that.

[Ernie was married, and had a small child at this time, which explains the urgency he had in getting a job amid the adverse economic conditions of the Great Depression.] "But, I was standing downtown in a public employment office on Third and Robert, and I was talking to another young guy about jobs and this and that. 'If you want a job,' he says, 'if you don't mind what kind of work you do, you can go to Seeger Refrigerator. They're hiring everybody they can get. It's a hell of a place to work; I would not work there myself.'

"You know, the next morning I was out to Seeger early in the morning. Yes, I got a job right away and they send you up on the line. They had a funny system—it was assembly line work. It was all divided into groups—groups of maybe 10 men to 30—all depends on the number of operations they had to perform. Each group was allowed so much money for so much work that they did. You did not get paid by the hour—it was all how much work you did on this cabinet. Each group was assigned a certain amount and then whatever money that group earned was divided amongst the men in the group. Then each group had what they called a group leader or a boss, and he would figure how many men he needed to perform the number of operations on the cabinets as they came down the line. You had to wait until one of these groups needed you and they would check you in. You'd have to stand along the wall in the morning—you didn't get paid for that! Maybe you'd stand there until noon or something, and then they'd say, 'Well, we won't need anymore today. Then you can go home.'

"So, I stood there and once in a while I'd get checked into a group—you know, work like hell and then go home. For the first two weeks I got a check for $24. Now after that they were starting a new cabinet and things weren't going very fast—so we did not make much money. After a while when I finally got a regular job on it, then we averaged $35 every two weeks, which wasn't very good, but anything helped to get by. So then I worked there for three, four years, or something like that. Then the war came along."

Ernst John Kulenkamp (1908–2005)*

*From an oral history made by Ernest John Kulenkamp (circa 1987). Reproduced by courtesy of the Kulenkamp family.

corners were rounded and the refrigerator's front sill was slightly raised to make cleaning easy. The porcelain lining of the refrigerator had one ground and two white coats of porcelain fused on 20-gauge sheet iron. This was the purest iron in commercial use. Once the linings were ready, they were fitted into the assembled cabinets through the top.

Seeger engineers conducted exhaustive tests before they selected the type of insulation that was used in *Coldspot* cabinets. This insulation, which was known as Dry-Zero, when formed into slabs and covered with waterproof paper, had the highest resistance to moisture absorption, rot, and mold that the engineers had found. These slabs, made to exact dimensions, were fitted snugly into all sides of the cabinet between the inner lining and exte-

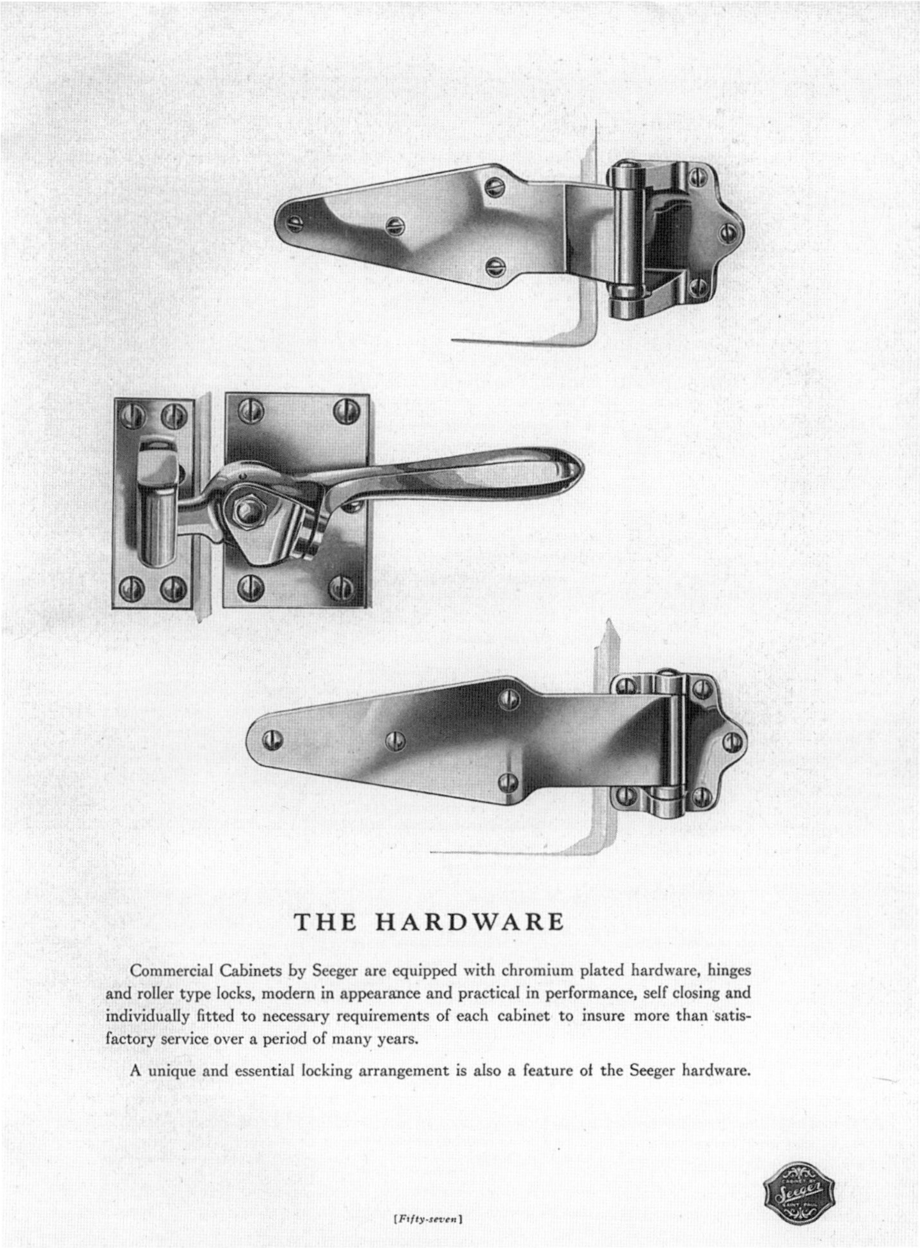
THE HARDWARE

Commercial Cabinets by Seeger are equipped with chromium plated hardware, hinges and roller type locks, modern in appearance and practical in performance, self closing and individually fitted to necessary requirements of each cabinet to insure more than satisfactory service over a period of many years.

A unique and essential locking arrangement is also a feature of the Seeger hardware.

[Fifty-seven]

When advertising its commercial refrigerator cabinets, the Seeger Company emphasized the quality of the hardware that it installed on these appliances because commercial buyers wanted assurance of reliability and durability in the refrigerators they purchased. From a Seeger catalog in the author's collection.

rior case. All joints were then sealed tight with hot hydrolene, thus sealing them against infiltration of air and moisture. Three inches of high-heat-resisting insulation was used in the six- and seven-cubic-foot *Coldspot* cabinets. Two inches of this Dry-Zero insulation was used in the four-cubic-foot cabinet.

Wood framing in and around the door opening served as a support for the refrigerator's hardware and trim. This frame was made of seasoned spruce and magnolia. The latter wood held screws solidly and was also particularly free from the twisting and cracking that was common in many woods.

Seeger Company advertising in the 1930s sometimes included lists of the various celebrities and famous people who had purchased refrigerators that had Seeger cabinets. This ad, for example, included Hollywood movie stars Norma Talmadge, Buster Keaton, and Gloria Swanson, industrialists Alfred I. and Eugene DuPont, the doctors William and Charles Mayo of the Mayo Clinic, and a number of British aristocrats. From a Seeger advertising booklet in the author's collection.

In the next step in assembly, a wood collar was pressed down on top of the insulation and liner. The sides, back panel, and liner were attached to the collar, which also helped to support the unit. An air-cushion rubber gasket sealed the joint between the cabinet and door unit. Sound-absorbing paper lined the door unit compartment. Air-cushion rubber gaskets made a tight joint at the door. The breaker strip or trim around the door and door opening was Bakelite, a heat insulating material, which prevented the transfer of heat into the liner. The Dulux finish was a very recent development of the DuPont Company. This finish was so tough that when applied to metal, bending, twisting, or even the severe blows of a hammer showed no effects. The usual kitchen stains and grease could be instantly removed with ordinary soap and water. Under its license from DuPont, Seeger engineered special equipment, developed scientifically controlled ovens, and trained skilled workers to produce this new finish on its refrigerator liners with reliable consistency.

First, the metal was finished smooth and thoroughly cleaned and dried. Then a primer coat was sprayed on, and baked 43 minutes at 390° F. Next, a white coat was sprayed and baked 43 minutes at 390° F. This coat, after baking, was carefully rubbed to remove any trace of unevenness. The third and finishing coat was then sprayed on and baked 86 minutes at 280° F. Unusual care was exercised in each spraying operation because just the exact amount of Dulux had to be applied to insure perfect finish. All operations of cleaning, drying, spraying, and baking were done on a chain conveyor traveling five and a half feet per minute. The Dulux process took eight and one-half hours from sheet metal shop to assembly line.

These 1937 photos capture Seeger people at work. From a Seeger News *in the author's collection.*

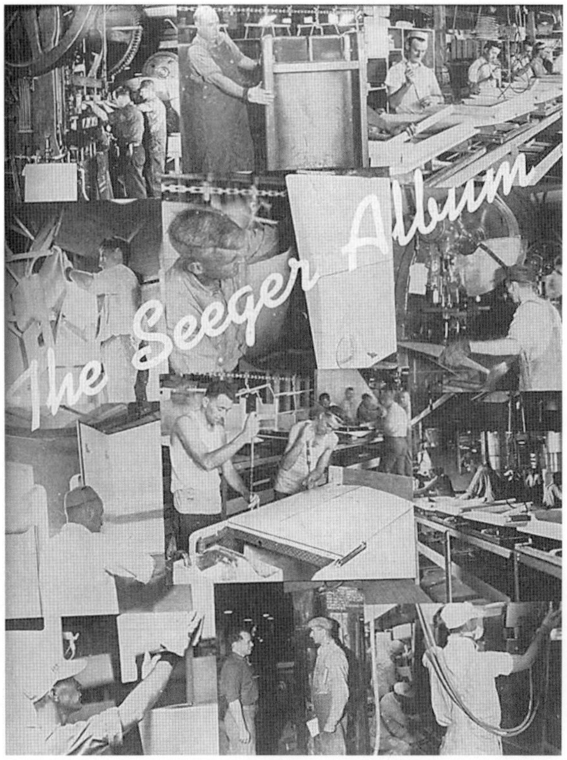

The all-porcelain models of Seeger refrigerators differed from Seeger's Dulux models in that more wood framing was used, and the outside shell was constructed on panels. Fusing glass on iron produces a porcelain enamel finish on a panel or part. The parts to be porcelainized first had to be cleaned carefully using pickling, cleaner, acid, and neutralizing baths. The material applied to the metal was a synthetic product ground in water and sprayed on with a gun. Soda ash, borax, nickel-oxide, manganese, feldspar, clay, silica, and cryolite were some of the materials used to produce the ground coat of "frit." This was a granular substance formed by fusing the ingredients in a furnace at 2,000° F, shattering

Voices *from the* Plant Floor *at* Seeger

The Special Room

"As a recent graduate of Cretin High School and a resident of the East Side neighborhood near Earl and Case [streets], I worked at Seeger for a few months about 1937. I was a runner in the Special Room, the center for filling orders from customers for replacement parts for their refrigerators. My task was to locate the part in any one of the factory's several departments such as the inventory of parts for its products, the Tool and Die shop if it was necessary for the part to be made, or the paint shop if it needed to be painted. We had lots of requests for replacement parts for the commercial refrigerator cases that Seeger manufactured. We worked hard and were kept very busy."

John L. Murray Jr.

John L. Murray Jr. Photo courtesy of the family of John L. Murray Jr.

it in cold water, then drying and grinding the mix. Finish coat frits were given different colors by the substitution of other materials for nickel-oxide. The basic frit was then mixed with other ingredients and ground with 40 percent water for ten to fifteen hours. The fineness was checked, and it was then allowed to stand for several days before using.

When the mixture was sprayed on the panels and parts, the water evaporated and left a fine, powdery residue. The panels were then put into a furnace and burned for four minutes at 1550° F for the ground coat. They were next sprayed with a white coat and passed through an oven to dry. The white was then wiped off at the edges and corners, allowing the tougher, black ground coat to show through. The second burning was for two minutes at 1,500° F. The finish coat of white was applied the same as the second, except it burned for two minutes at 1,480° F.

By 1936, production of refrigerators at Arcade Street had reached occasional peak volumes of one hundred or more cabinets per hour in periods of heavy demand. Seeger employees were producing a vastly increased number of quality refrigerators for domestic and commercial use.[3] A year later after the Seeger Company began manufacturing refrigerators for Montgomery Ward, the company shipped 308,551 refrigerators from its Arcade Street plant.[4] The residential and commercial demand for refrigeration units in the late 1930s forced Seeger to expand its facilities to serve the manufacturing needs of a factory employing 2,000 workers and operating with three shifts twenty-four hours a day.

This second expansion program began in 1936 when a three-story building was erected as an addition to the original building. This new building allowed Seeger engineers to increase the floor and operating space of the overtaxed Metal Shop. This expansion of the Metal Shop filled the first two floors of the new building. All main office operations were then centralized on the new building's third floor. Shortly after this new building was completed, the Seeger plant crossed the city street that passed its doors and began to grow even more with the construction of a large, one-story, fireproof building that was designed for the warehousing of finished goods.

The company's thirty-sixth year of growth was marked by a six-month period of decreased production and great personal loss when on August 28, 1938, the venerable business and community leader, John A. Seeger died at

Seeger employees had a marching club that participated in the 1938 St. Paul Winter Carnival. The carnival outfits consisted of a scarlet red-bodied jacket with dark gray sleeves and ski pants topped off by scarlet and gray pullover caps. The script Seeger logo on the jackets had blue lettering. This photo was taken with the unit standing in front of an ice castle. Photo from the author's collection.

the age of eighty-three. Despite this great loss, the Seeger Company continued vigorous operation under the leadership of the founder's sons, Walter and Gus Seeger.

The factory on the East Side of St. Paul was now one of the largest producers of refrigerator cabinets in the country. It began producing refrigerator cabinets as never before. Lights burned through the night, another floor was added to the new warehouse, and as production continued at peak capacity on and through 1940, planning began for another expansion program only to be stopped in the discussion stage by the turmoil and uncertainty that characterized a world that seemed to be drifting toward war in Europe. Uncertainty and tension gripped the nation. Drastic laws were passed, as the U.S. government desperately prepared for the inevitable, by regulating the lives of men and machines. In 1942 the overriding requirements of national security curtailed refrigerator production and brought to a stop the organization that for thirty-four years had lived and produced within the boundaries of a single product.

At the outbreak of World War II, the factory's assembly lines were

divided into two facilities: the Commercial Assembly Line and the Domestic Assembly Line.[5] Each was a separate department in itself, in that the Commercial Assembly Line was dedicated to the assembly of Commercial Display Cases and Special Coolers. The Domestic Assembly Line was responsible for the assembly of household refrigerators. The interesting feature of these departments was that an observer could start at the beginning of the assembly line and watch the various parts assembled and in the course of about 50 minutes see a refrigerator completely assembled and crated for shipment. Or, that observer could stand at the inspection point and watch about 125 completed domestic refrigerators pass by in one hour, or go to the inspection point on the Commercial Assembly Line and see at least five completed commercial display cases go by in one hour.

The Domestic Assembly Line was the second largest department in the plant (after the Engineering Department) with a floor space of approximately 718,000 square feet, and under full refrigerator production had a staff of employees of about 150 people on each of the three shifts. The assembly line was 700 feet long and could produce up to 125 completed cabinets per hour of the smaller sizes of refrigerators, or 1,000 completed refrigerators in an eight-hour shift.

John A. Seeger and Elvina Yoerg Seeger were married in 1877. These are their wedding portraits. In the lower photo from about 1935, John A. and Elvina Yoerg Seeger posed for a photo in their home at 657 Fifth Street East in St. Paul. The Seegers lived in this house from the time it was built about 1885–86 until their deaths. John died in 1938 and Elvina in 1940. The Seeger wedding portraits are courtesy of Miriam S. Bell and the 1935 photo of the Seegers is courtesy of John S. Reay.

8

1942–1945: A Partner in National Defense with Wartime Production

THE OUTBREAK OF WORLD WAR II brought a freezing order on the manufacture of refrigerators.[1] Nevertheless there still was a demand for refrigerators in the United States. To fill this need, Seeger assembled two wartime home appliances. The first was a new product based on an old procedure, which Seeger manufactured for Sears and Montgomery Ward. It was the Ice Conversion Model of refrigerator with a separate compartment built into the cabinet for holding ice. Because ice took the place of a mechanical unit in this compartment, the Ice Conversion Model was similar in its features to the old style icebox. After the war, when electrical units could be obtained, the ice compartment unit was easily removed and replaced with a compressor. Second, the War Production Board of the federal government also allowed Seeger to produce the Victory Refrigerator. This model was built with less than six pounds of metal in the cabinet. Prewar electric refrigerator cabinets required as much as two hundred to three hundred pounds of steel or other metals. The wartime appliance was constructed of Masonite or other pressed woods, and when properly constructed was very efficient. All of the cabinets made by Seeger had the same high standard of workmanship as their prewar metal refrigerators.

After December 7, 1941, Seeger quickly shifted from manufacturing refrigerators to the production of a variety of defense products. The resources of the Seeger factory were totally committed to producing materials essential for the war effort. As the new situation brought new demands, changes were inevitable. The most important was Seeger's first intentional recruitment of women for its labor force. Some women had been employed at Seeger before the war, but the specific number cannot be known due to the destruction of these personnel records. Up to 1942 the company had made no effort to recruit women. With many of the men who had comprised the Seeger workforce now serving in the armed forces coupled with a definite shortage of qualified men who for whatever reason were not in uniform and available to

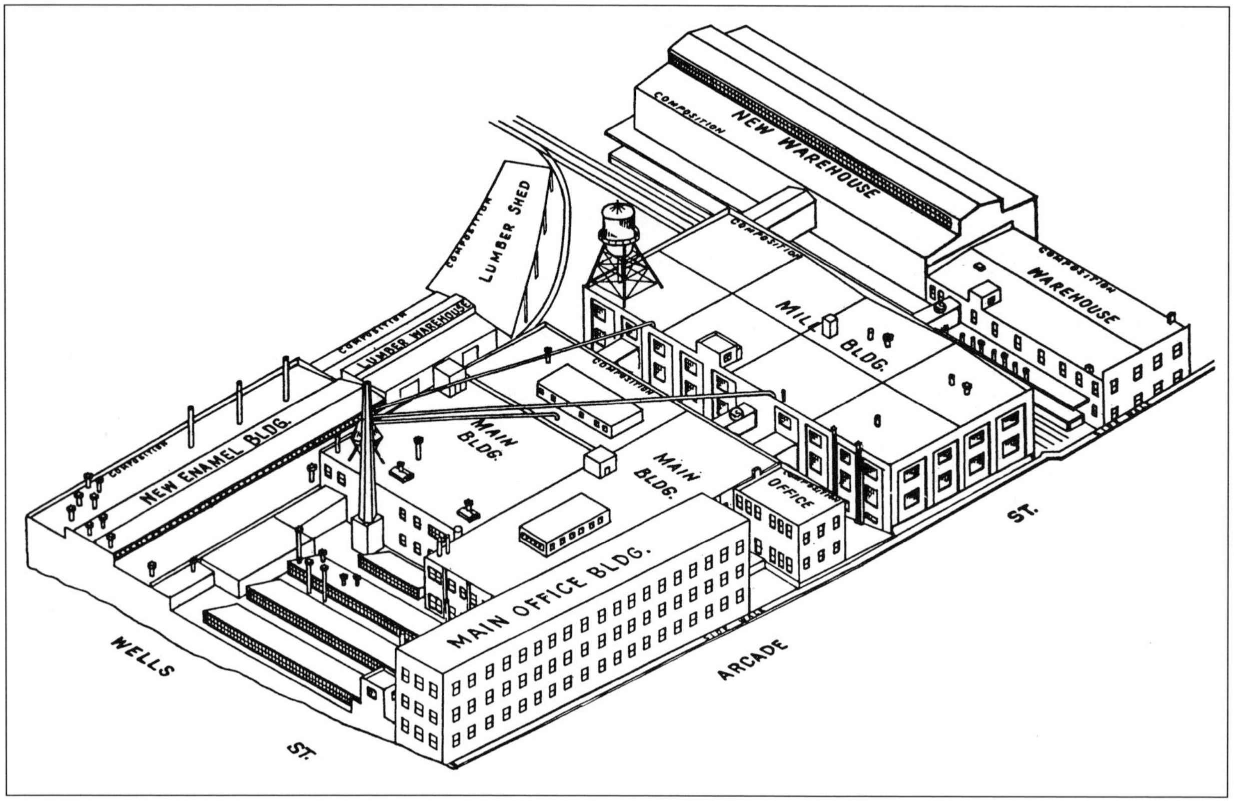

This diagram from the late 1930s shows the layout of the Seeger Refrigerator Company and its major office, manufacturing, and warehousing areas. From a brochure in the author's collection.

join the Seeger ranks, women joined the departments and assembly lines of the company. Sadly there are no records of how many women began working at Seeger during the war. All the evidence that is available is either anecdotal or very fragmentary.

In support of the war effort, Seeger produced ammunition boxes that were all steel and gas-welded, spot-welded, and projection-welded. They were used for storage of ammunition belts for use on machine guns aboard ships. In addition the company manufactured bomb racks for the B-29 Superfortress bomber, M-26 aircraft parachute flares, M9A1 rifle grenades, M20A1 boosters, 45-caliber shell cases, flexible ammunition shutting, .30- and .50-caliber, and 20- and 205-millimeter shell containers, 90-millimeter shell cases, high explosive M71 shell cases, MKVI .50 ammunition boxed, naval rocket nozzles, bazooka nozzles, Oerlicon lockers, V. S. No. 6 & V. S. No. 7 refrigerators for the U.S. Navy, 96-foot coolers, and assorted precision aircraft parts.

In the spring of 1942, Northwest Airlines of St. Paul signed a contract with the Seeger Experimental Department to begin to manufacture a va-

In these late 1930s photos, Gustave R. Seeger (left), secretary and treasurer, reads his mail at his desk and Walter G. Seeger (right), president of the Seeger Company, dictates a letter. The Great Northern Railroad seal on the wall behind Gustave Seeger reflected Walter Seeger's service on the railroad's board of directors. Both photos are from the author's collection.

riety of bomber modification parts for the B-24 Liberator bomber that Consolidated Aircraft had developed for the U.S. Air Force in the late 1930s. By the end of the war, more B-24s had been produced than any other American combat aircraft. In the modification of the B-24, feeding ammunition to the swivel machine guns in the plane's various turrets was a difficult problem. During the bomber's development, a flexible ammunition feed chute had been made for this purpose, but like other items needed for a global war, these chutes were too little and too late.

Beginning in 1942 the Seeger Company produced a variety of ammunition boxes for the U.S. military. In this photo the shell of the all-steel box is being spot-welded to the box's front panel. From a Seeger News *in the author's collection.*

On a Sunday afternoon, in August 2, 1942, Warren "Nobby" Nobles, Seeger's Assistant Chief Engineer, was checking over the list of parts that Northwest Airlines had ordered from Seeger. Northwest had expressed an urgent need for some type of flexible ammunition feed chute that could be easily adapted to all types of planes and could be produced as fast as needed. Wanting to do something about this problem, "Nobby" went to work in his basement workshop, and by three o'clock the following morning, he had a sample chute design worked out. The next morning work was started in the

Seeger Refrigerators and U.S. Navy Minesweepers

Seeger supplied defense products to the U.S. Army and Navy during World War I. After December 7, 1941, the company's manufacturing operations were speedily transformed and again began to build a number of products for the U.S. forces. By early 1942 Seeger was producing refrigerators for minesweepers built in numerous U. S. shipyards for the American and British navies. At least 274 U.S., warships and 49 British vessels were outfitted with Seeger refrigerators (see Figure 1).* Seeger's Engineering Department's detailed plan of the appliance, Figure 2, notes in precise detail the specifications, materials, and size of the refrigerator. The wiring for the shipboard appliance is noted in Figure 3.

In addition to the refrigerator, Seeger manufactured an oak locker that held spare parts for the appliance. Figure 4 describes the specifications for the construction of the thirty-pound box that was to hold the 11.5 pounds of refrigerator parts (Figure 5) should the appliance fail to operate properly at sea.

***_Instruction Book for Refrigerators on YMS 135-409, BYMS 31-80, Manufactured by the Seeger Refrigerator Company_ (St. Paul, 1942). During World War II, the designation "YMS" meant "motor minesweeper." These small, wooden-hulled vessels, depending on their class, displaced between 245 and 320 tons, had a length of 136 feet, a beam of 24.5 feet, and a draft of 10 feet. Most of them were built in 1943 and 1944. They operated in the Atlantic, Caribbean, and Pacific Theatres. With a crew of thirty-two officers and men, a naval lieutenant was typically the commanding officer. These minesweepers had a main battery of one 3" gun and two 20mm guns. They also carried two .50-caliber machine guns and two depth charge racks. In addition to minesweeping, they also served as escorts on convoy duty, carried out antisubmarine patrols, and provided defensive fire against Japan kamikazes in the final assault on Japanese-held islands in the Western Pacific. A number of these minesweeper participated in the landings at Normandy, France, on D-Day (June 6, 1944), and one of them was damaged in the attack.**

Experimental Department to build a five-foot sample of a flexible ammunition chute for .50-caliber ammunition. By August 14th the sample chute was completed and submitted to Northwest Airlines for an actual firing test, which was first made from a waist gun on a B-24 over Lake Superior. This test proved satisfactory. Actual firing tests were then made at Wright Field in Dayton, Ohio, and on their approval the development of the chute decreased to .30-caliber and 20-millimeter chutes, and the numerous fittings and attachments needed for these various weapon systems.

The chute itself consisted of stamped strips of stainless steel designed in such a way that each piece weaved into and interlocked with the next in an ingenious manner. The flexibility and adaptability of "Nobby" 's design to many sizes made this chute almost indispensable. Soon air force and other military procurement commands and aircraft manufacturers throughout the United States accepted the Seeger Flexible Ammunition Chute. Immediately the Seeger factory began producing the Flexible Ammunition Chute in various sizes, literally providing a way to help "Pass the Ammunition."

Parachute Flares for the Army

Seeger also manufactured parachute flares for the air force on missions of observation, reconnaissance, and bombardment. Parachute flares were a modern weapon of war and used successfully in all theatres of World War II. The U.S. air force used them to illuminate enemy targets on their daily long-range bombardment missions to minimize reinforcement of Axis troops on the ground.

The name "parachute flare" was derived from the method in which the flare was operated. The flare's intricate assembly of parts was encased in a metal cylinder that resembled a bomb in its outward appearance. In operation, the flare was dropped from a plane flying at altitudes as high as 25,000 feet, and at speeds of 200 to 300 miles per hour. As the flare dropped toward the earth, a series of automatic and timed explosions of various fuses occurred, the parachute, similar to that para-

INSTRUCTION BOOK
FOR
REFRIGERATOR
ON
YMS 135-409
BYMS 31-80

WHEELER Y M S PURCHASING COMPANY P. O. NO. C-20

MANUFACTURED BY
SEEGER REFRIGERATOR CO.
St. Paul, Minn.

1942

ALLOCATION OF ORDER BY SHIPBUILDERS

SHIPBUILDER	CITY and STATE	CONTRACT NO.	BOAT NO.
AMERICAN CAR & FOUNDRY	WILMINGTON, DELAWARE	DA 501	BYMS 31-36
ASSOCIATED SHIPBUILDERS	SEATTLE, WASHINGTON	DA 145	YMS 287-296
ASTORIA MARINE CONST. CO.	ASTORIA, OREGON	DA 502	YMS 135-142
BALLARD MARINE RY. CO. INC.	SEATTLE, WASHINGTON	DA 503	YMS 326-333
BARBOUR BOAT WORKS	NEW BERN, NORTH CAROLINA	DA 504	BYMS 37-42
BELLINGHAM MARINE RY. & B. B. CO.	BELLINGHAM, WASHINGTON	DA 505	YMS 269-276, 342-345
BURGER BOAT COMPANY	MANITOWOC, WISCONSIN	DA 506	YMS 155-162
CAMPBELL MACHINE COMPANY	SAN DIEGO, CALIFORNIA	DA 141	YMS 151-154
DACHEL CARTER S. B. CORP.	BENTON HARBOR, MICHIGAN	DA 508	YMS 163-170
GREBE, HENRY C. CO.	CHICAGO, ILLINOIS	DA 510	YMS 171-182, 279-280, 405-409
HARBOR BOAT BUILDING CO.	TERMINAL ISLAND, CALIFORNIA	DA 155	YMS 313-316, 303-306
HILTEBRANT, C. DRYDOCK CO.	KINGSTON, NEW YORK	DA 512	YMS 195-206
KRUSE & BANKS S. B. CO., INC.	NORTH BEND, OREGON	DA 115	YMS 265-268
LARSON A. BOAT SHOP, INC.	TERMINAL ISLAND, CALIFORNIA	DA 514	YMS 320-325
MARTINAC, J. S. B. CORP.	TACOMA, WASHINGTON	DA 515	YMS 216-221, 277-278
MOJEAN & ERICSON	TACOMA, WASHINGTON	DA 516	YMS 222-225
NORTHWESTERN S. B. CO.	SO. BELLINGHAM, WASHINGTON	DA 143	YMS 285-286
RICE BROS. CORP.	E. BOOTH BAY, MAINE	DA 517	YMS 303-312
SAN DIEGO MARINE CONST. CO.	SAN DIEGO, CALIFORNIA	DA 140	YMS 143-146, 281-284
SEATTLE S. B. & D. D. COMPANY	SEATTLE, WASHINGTON	DA 519	YMS 334-341
SOUTH COAST COMPANY	NEWPORT BEACH, CALIFORNIA	DA 114	YMS 259-264, 317-319
STADIUM BOAT WORKS	CLEVELAND, OHIO	DA 520	YMS 235-240, 389-392
TACOMA BOAT BUILDING CO.	TACOMA, WASHINGTON	DA 521	YMS 241-246, 297-298
WEAVER SHIPYARDS	ORANGE, TEXAS	DA 522	YMS 247-258, 371-374
WESTERGARD BOAT WORKS OF BILOXI.	BILOXI, MISSISSIPPI	DA 523	BYMS 55-64
WESTERN BOAT BUILDING CO.	TACOMA, WASHINGTON	DA 524	YMS 147-150
WHEELER SHIP BUILDING CORP.	WHITESTONE, NEW YORK	DA 525	BYMS 65-80, YMS 363-370
NEVINS, HENRY B. INC.	CITY ISLAND, NEW YORK	DA 188	YMS 397-404

33854

Figure 1

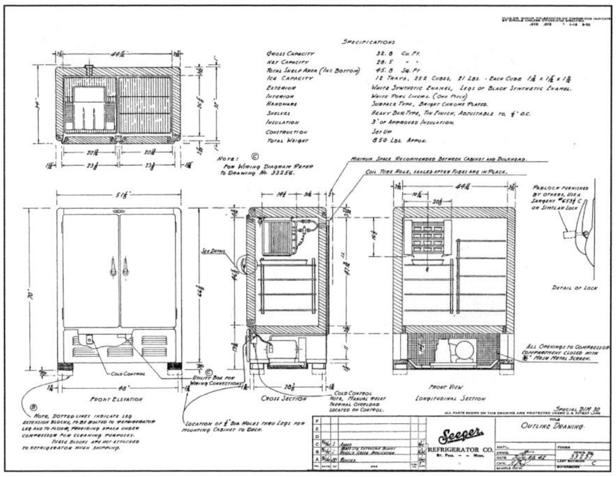

Figure 2

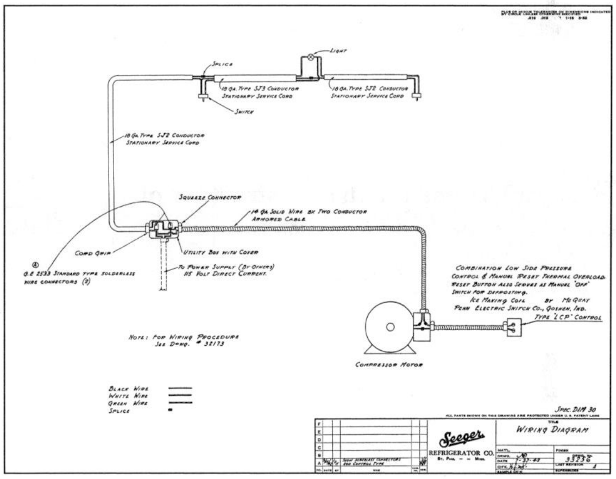

Figure 3

LIST OF SPARE PARTS AND TOOLS

NAVY MINE SWEEPER — ELECTRIC REFRIGERATOR MODEL DIM-30

NAME OF VESSEL__________ — 115 VOLT D.C. MOTOR

SEEGER REFRIGERATOR COMPANY

ITEM NO.	NO. OF SETS	NO. PER SET	NAME OF PARTS	UNIVERSAL COOLER CORP. PART NO.	SEEGER REFRIGERATOR CO. PART NO.
1	1	1	EXPANSION VALVE		33590
2	1	1	CONTROL	44527	33633
3	1	1	SEAL ASSEMBLY	1014	33634
4	1	1	SEAL PLATE	44478	33635
5	1	1	SEAL PLATE GASKET	41984	33636
6	2	1	BELTS	44629	33637
7	1	1	VALVE PLATE ASSEMBLY	62641	33638
8	1	1	VALVE PLATE GASKET	43369	33639
9	1	1	CYLINDER HEAD GASKET	42477	33641
10	1	1	S.O.V. GASKET	11247	33642
11	1	1	SUCTION CHAMBER GASKET	42475	33643
12	1	1	BOTTOM PLATE GASKET	41995	33644
13	2	1	DEHYDRATOR	82451	33645
14	1	1	PRESSURE GAUGE	10544	33646
15	1	1	COMPOUND GAUGE	5172	33647
16	1	1	GAUGE FITTING	80115	33648
17	1	1	RATCHET WRENCH	5167	33649
18	1	1	CHARGING LINE		33652
19	1	1	1/8 PT. x 1/4" SAE FITTING	5117	33653
20	2	1	MOTOR BRUSHES		33654
21	1	1	THERMOL HEATER COIL		33655

PURCHASE ORDER No. C-20
SHIPPING ORDER No. 11710
33848

Figure 5

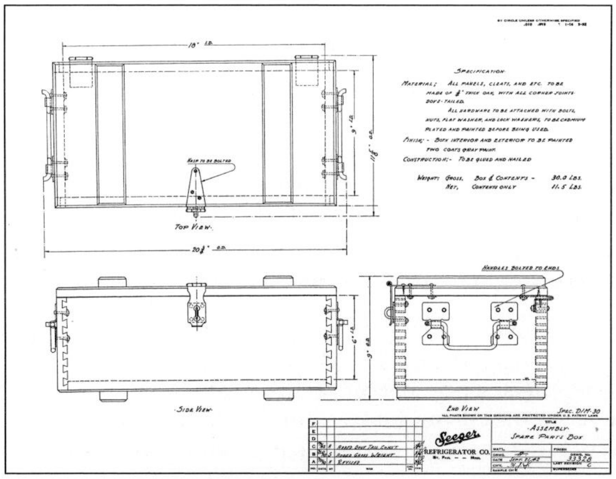

Figure 4

From a U.S. Navy instruction book in the author's collection.

On March 17, 1942, the Seeger Company ceased production of household refrigerators in compliance with U.S. government directives. This was the last household refrigerator produced in the plant for the duration of the war. Full production of defense-related products followed. From a Seeger News *in the author's collection.*

chute used by the aviator, was ejected from the case of the flare, and in turn, the flare assembly or the candle and deflector shade were released and held suspended from the parachute. The flare was then automatically lighted and developed a minimum of 800,000-candle power and burned for 3 to 3 and-a-half minutes. The flare's complex operational mechanism required very careful and precise assembly. There were several hundred different parts and assemblies in each flare and all had to be made to exact dimensions.

Bomb Racks for Army Air Force B-29 Superfortresses

In December 1942 the Seeger Company received a contract from the Bell Aircraft Corporation of Marietta, Georgia, and the Boeing Aircraft Company of Wichita, Kansas, to manufacture parts for the bomb rack in the B-29 Superfortress. Boeing had developed the B-29 for bombing Japan from long distances. It and Bell Aircraft were the principal manufacturers of this long-range bomber that was only used in the Pacific Theatre during World War II. The B-29 made its maiden flight on September 21, 1942. Thus the contract for building the bomb racks that the Seeger Company received some two to three months later was only a small, but necessary, part of an immense aircraft building program.

Seeger engineers took great care in building the jigs, measuring gauges, and in procuring the

Voices *from the* **Plant Floor** *at* **Seeger**

He Worked His Way Up to Head of Inspection

"I began working at Segeer's during the summers, about 1927. I attended Mechanic Arts High School and graduated from the St. Paul Vocational School in 1930 as a cabinetmaker and the honor student of the year. I first worked at Seeger making commercial display cases for General Electric—putting their nameplate on the boxes. John A. Seeger would walk into the plant every morning and would always come over and talk to me. I worked in the Woodworking Department from 1933 to 1936, and became in charge of metal assembly—the spot welders. I was deferred during World War II to aid Seeger's production of wartime products. After the war I worked in Quality Control becoming head of the Inspection Department. I retired from the company in 1974."

Harry Gauld

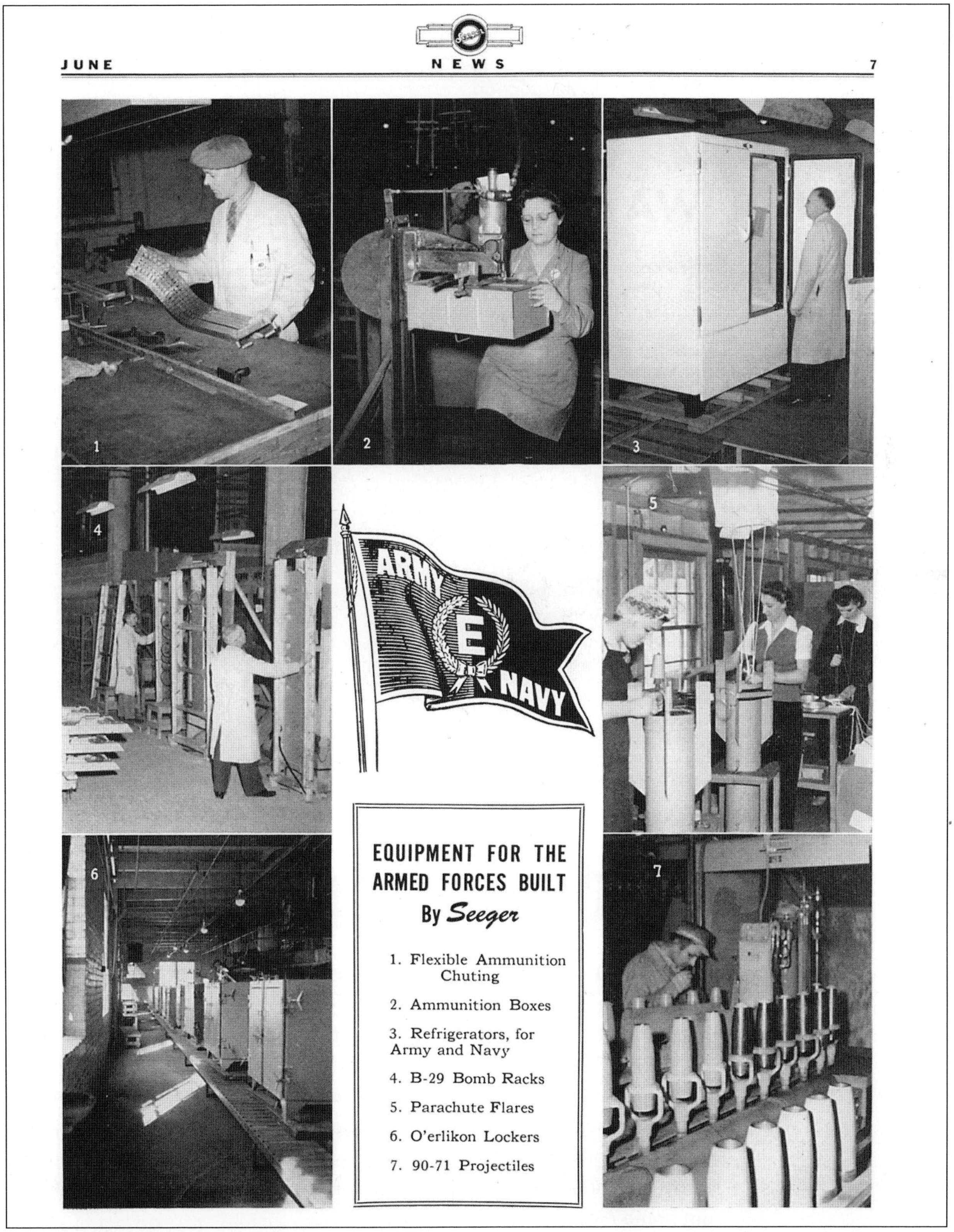

JUNE — SEEGER NEWS — 7

EQUIPMENT FOR THE ARMED FORCES BUILT By *Seeger*

1. Flexible Ammunition Chuting
2. Ammunition Boxes
3. Refrigerators, for Army and Navy
4. B-29 Bomb Racks
5. Parachute Flares
6. O'erlikon Lockers
7. 90-71 Projectiles

This page from a 1945 copy of the Seeger News *showed all the equipment that the Seeger Company manufactured in support of the war effort. In the center of the page is the Army-Navy "E" pennant that was awarded to the Seeger Company in 1944 for the high quality of the products the plant produced for the armed forces. From a* Seeger News *in the author's collection.*

The cover of the September 1944 issue of the Seeger News *displayed a photo of a B-29 Superfortress, a long-range air force bomber for which the Seeger Company built bomb racks. From a* Seeger News *in the author's collection.*

proper equipment so that the parts they would manufacture would meet the exacting requirements of Army Air Force specifications and to ensure that the bomb racks they built adhered to the close dimensional tolerances that were required. Production was started in the former Metal Assembly Department, but in a short time it was found that the department would be too small and plans were made to move into the new warehouse building. The first floor was cleaned out and the entire Bomb Rack Assembly department was moved over and production immediately was underway to turn out bomb racks in any quantity needed.

The bomb rack manufactured at Seeger for the Army Air Force's B-29

Superfortress was constructed entirely of aluminum. It required approximately 2,300 pounds of aluminum for one set of racks. As every part of each set had to be made to very close tolerances, it was necessary to build special jigs, fixtures, and even tools with which to produce racks that could pass thorough inspection. It also required training Seeger employees in jobs entirely foreign to their previous experience manufacturing refrigerators and freezers. The men and women who worked on this product realized that they were making a part that must function precisely and that the tiniest flaw might mean the difference between hitting and missing a target.

These women are bolting a mounting terminal to a rail that was part of the bomb rack assembly in the Superfortress. From a Seeger News *in the author's collection.*

On June 15, 1944, the long-awaited report that the new B-29 Superfortresses were in actual combat was flashed around the world. On that day forty-seven B-29's had left a base in China and made their first bombardment of Japan. Bomb racks were not simply racks in which bombs were stored; they were one of the most important parts of the Superfortress. For without a bomb rack, the plane would have been of little use as a bomber. Located inside the plane in what was known as the bomb bay, these ranks held the bombs that were released over enemy targets.

The bomb rack for a Superfortress was a precision-built piece of equipment. The contract's specifications required that each rack be built to dimensions within tolerance of .005 (five thousands of an inch.) Each rack had to be interchangeable with any other so that for varied bomb loads, racks could be replaced easily to accommodate whatever weight of bombs was required for each mission. There were sixteen racks to one set of bomb racks, each set of bomb racks required 14,002 individual parts to make up a complete set.

A complete set of bomb racks consisted of three individual sizes of racks. Each size rack was assembled on a separate assembly line. Each rack was progressively assembled as it was moved down the line from one operation to the next. Assembly fixtures of special design were used in each operation assuring accurate assembly of all

Here a team of men assembled several parts for a bomb rack in a master jig so that they could precisely drill a series of holes. From a Seeger News *in the author's collection.*

This photo shows the Seeger nameplate on one of the bomb racks in the restored B-29 Superfortress that is part of the collection at the New England Air Museum at Bradley International Airport, Windsor Locks, Conn. Photo courtesy of the New England Air Museum.

A second photo from the restored B-29 Superfortress shows how two (inert) bombs were mounted on the Seeger-built racks in the aircraft's bomb bay. Photo courtesy of the New England Air Museum.

parts. Parts were assembled into the various jigs, checked, and then drilled in position. After removing the parts from these jigs, the parts were carefully deburred, inspected, and forwarded to the next operation. Heat-treated rivets were stored in special Seeger-built rivet coolers that were maintained at a temperature of -10° F until used in a rack.

Each rack was carefully checked in production by the use of gauges as it moved down the assembly line. Inspectors continually checked all operations along the line by inspecting all major dimensions such as the spacing and operation of bomb hooks and bomb hook assemblies, the spacing of mounting terminals and fittings, the placing and soldering of all electrical connections, and the installation of fusing solenoids. Rivets were carefully checked for proper size and for possible cracked heads.

The completed bomb rack was then sent to the Final Inspection and Functional Test Department. Here each rack was placed into a test fixture that checked the alignment of the top mounting terminals and the lower mounting fittings. The rack was then placed into a functional or operation test fixture where it was tested for the proper performance of all electrical controls and the operation of all bomb release stations. This test was identical to the operation of the rack in actual use. Thus it included the firing of one, two, or all stations. Bomb shackles were tried on each rack to assure fit on hooks. Identification numbers and serial numbers were then placed on the racks. Army Air Force inspectors supervised the assembly of all bomb racks and on completion of each rack, the inspectors reviewed each rack and the AAF stamp of acceptance was placed on each serial number plate. The bomb racks were then carefully crated for shipment to prevent any possible damage to the racks or parts of them.

Voices *from the* **Plant Floor** *at* **Seeger**

The Company Was Good to Me

"I quit school at sixteen and went to Seeger's in the morning and started my job that afternoon. It was wartime and we were making bomb racks, flexible machine gun chutes, and bombing flares to light up the night.

"When I started at Seeger they were very strict in how a place should be run. One wouldn't dare try and have a cigarette during work hours, only at break time. As years went by, I would say it got out of hand. At the time it was tough working, but I didn't know anything different. I believe workers tried harder to do quality work than in later years.

"The company was good to me. I learned to spot weld, gas weld, metal finish, and run various machines. It paid off because my boss (Group Leader) could only weld. We had a cut in production and I got his job. I could do everything. I ended by becoming a foreman.

"I had a good life at Seeger-Whirlpool, raised seven children, and my wife stayed home and I sometimes worked two jobs to support them. And they all are doing well."

Clinton Conrad

Pressure Cookers

On the home front World War II had an impact on the daily life of every American consumer. The federal government curtailed the packing or canning of vegetables, fruits, and preserves for private or home consumption in order to conserve the vitally needed tin that was used for tin cans. This situation required individuals and families to do most of the preserving at home. The most economical and common method of preserving and canning that was then available was known as "pressure cooking." This prewar cooker was made of aluminum, but as aluminum was another much needed war material, even pressure cookers could no longer be made of aluminum. Because steel was more readily available than aluminum, the government approved and allocated the necessary steel to make the new pressure cooker for the canning of food.

This new pressure cooker was of the same design and principle of operation as the aluminum cooker, but all the parts were made of steel and then given a porcelain enamel coating. The Seeger Company made the "kettle" part for this cooker under a subcontract from an Eau Claire, Wisconsin manufacturer. The kettle was drawn to shape in the Press Room in the Metal Shop, an operation that required a number of draw-dies and required many

Here workers in the Seeger plant used a series of draw-dies positioned in a big press in the Metal Shop to shape the kettle portion of a pressure cooker. From a Seeger News *in the author's collection.*

operations to complete in the shape required. After all Metal Shop operations were completed, the kettle was moved to the Porcelain Enamel Plant where the kettle was put through a process of enameling that required scientific control. The kettle had to be "fired" at certain temperatures that would control the baking of the enamel and would not warp the kettle. All parts were carefully handled and cleaned before starting through the process of applying the enamel to the entire surface. After the kettle had been sprayed, fired, or baked, it was carefully inspected and gauged for the size, flatness, and fit of the cover. The kettles that passed inspection were then packed into individual cartons for shipment to another concern for final assembly.

The "E" Award

In recognition of its unstinting wartime production efforts, the Seeger Company was presented with the Army-Navy "E" Award in a colorful ceremony the afternoon of October 21, 1944. Grouped before a flag- and bunting-draped platform on which the flags of the United Nations were displayed, Seeger employees looked on as Lieutenant Colonel Bailey A. Wright, the

Area Representative for the Mid-Central Procurement District of the Air Technical Service Command of the Army Air Forces presented the Army-Navy "E" pennant to Walter G. Seeger, president of the company. Following the presentation of the "E" pennant, Commander George E. Jacobs, Inspector of Naval Material for the area of the country that included St. Paul, presented the Army-Navy "E" lapel pin to Mr. Seeger.

Walter G. Seeger, president of the Seeger Refrigerator Company, received the Army-Navy "E" pennant on October 21, 1944, for the high quality of production work that was done in the Seeger plant in support of the war effort. Commander George F. Jacobs, U.S. Navy (left) and Lieutenant Colonel Bailey A. Wright, U.S. Army made the presentation to Mr. Seeger and all the employees of the Seeger Company. From a Seeger News *in the author's collection.*

In 1906 the Navy had instituted an award to be known as the Navy "E," for excellence in gunnery. This award was later extended to include outstanding performances in engineering and communications. An honor not easily won or lightly bestowed, it became and remained a matter of deep pride to all who received it.

Following the attack on Pearl Harbor, the demand for war production exceeded anything the United States had ever known. The general public soon became aware that our fighting forces and the men and women of industry were partners in this great struggle to maintain human freedom.

After the award ceremony, Walter Seeger (right center in the civilian suit) and Gustave Seeger (left center with patterned necktie) led their military guests on a tour of the Seeger plant. From a Seeger News *in the author's collection.*

Inspection of the military products manufactured at the Seeger plant was very rigorous. Principal Inspector F.S. Bradford (left to right); then Lieutenant Commander George F. Jacobs, U.S. Navy; Commander B.K. Culver, U.S. Navy; and Resident Inspector George Armstrong oversaw inspection of all materials produced for the navy. From a Seeger News *in the author's collection.*

To promote this awareness and to inspire all men and women of industry to do their utmost in producing the vital material and equipment needed by the armed forces, the U.S. Army and Navy cooperated in honoring all industrial workers in war production with a combined symbolic "E" for Excellence. The award came from the combined Army and Navy boards of recognition as a citation for work achieved and as an inspiration for ever-increased production.

The Army-Navy "E" was awarded to plants for the high quality and quantity of their production of war material and that had also complied with other important factors that affected war production. These criteria included full utilization of available equipment; avoidance of work stoppages; maintenance of fair labor standards; cooperation with the war program; effective management and engineering; satisfactory maintenance of records on accidents, working conditions, safety, and plant protection or security.

The coveted Army-Navy "E" was awarded to the Seeger Refrigerator Company at the request of the Technical Service Command of the Army Air Forces for outstanding production of army and navy equipment. This Army-Navy "E" pennant was the highest award a company could receive and it signified that the employees of the Seeger Refrigerator Company were doing their utmost to help the nation's armed forces to victory.

Jet Fuel Tanks

After World War II had ended, Seeger resumed the manufacture of domestic and commercial refrigerators and freezers, but during the years of the Korean War (1950–53), the company successfully combined the production of these prewar products with defense materials. The principle wartime contract was for the manufacture of jet fuel tanks for the air force's jet fighter planes. A jet fighter used such vast quantities of fuel that it had to be fitted with auxiliary fuel tanks if it was to travel far from its base. Beginning in April 1951, these auxiliary tanks were made at the St. Paul plant in the same area, Building 17, that had been used during World War II to produce 90-millimeter shells and the B-29 bomb racks. Nearly one hundred people at Seeger were engaged in the precision production of fuel tanks under a subcontract with North American Aviation Inc.

As had been the case with bomb racks a few years earlier, the Korean War contract required precision production work. Any irregularities in the contour of a suspended auxiliary fuel tank would affect the flight characteristics of the jet plane to which the tank was secured because the plane moved at such a high speed that the air through which the aircraft passed acted more like a liquid than the gas it really was. These tanks were made of sheet aluminum that was carefully formed and welded into a hollow, streamlined shape.

The production facility was composed of five enclosed areas. Seeger inspectors used one of these areas and the others were set up for salvage, quality control, production control and the timekeeper, and for a tool room. These "cages" were built in one day and painted the next. Lockers were then installed in a locker area and welding units were set up so that the first "built-by-Seeger" fuel tank was completed on June 8, 1951. Each pair of the wing tanks that the plant produced gave greater range to a U.S. jet fighter. After being formed into the desired shape, the aluminum tank shell parts were carried by a conveyor to a workstation where they were sheared to their finished size on routing fixtures. Following welding of the parts into the streamlined shape specified in the blueprints, each tank was carefully inspected prior to being readied for shipment to the air force.

Seeger, unlike many corporations, did not reap any windfall profits

In the late 1940s Seeger produced jet fuel tanks, such as the one seen here, for the U.S. Air Force. Standing beside this military hardware is Beverly Wingard, who at the time this photo was taken was a candidate for the position of Queen of the Snows at the St. Paul Winter Carnival. The Seeger Company was her sponsor for this civic festival. The photo is in the author's collection.

With the cessation of hostilities in the fall of 1945, the Seeger plant returned to manufacturing household refrigerators. Here workers on the assembly line installed a gasket around the door opening of a refrigerator shell. From a Seeger News *in the author's collection.*

from its wartime defense production contracts. As noted in Appendix 3, the company's revenues during the war years remained a constant percentage of between 5 and 5.5 percent of its gross annual sales. Many Seeger employees were called to serve in the armed forces and a number of them were wounded or died in combat. As a corporate citizen, Seeger, was doubtlessly driven by a sense of patriotic commitment to relentlessly serve the national war effort. To generate a windfall profit on operations under such circumstances of a national emergency would have been unpatriotic.

9

1945–1955: A Decade of Robust Productivity

With the close of World War II in Europe in May 1945, executives at the Seeger Company began to consider the firm's needs and strategies to serve a postwar marketplace. The company searched for new ways to produce its goods and survive in an expanding and increasingly competitive world. The first step in this direction was the announcement made on May 12, 1945, that Seeger would acquire the Sunbeam Electric Manufacturing Company of Evansville, Indiana, and that the new company would be known as the Seeger-Sunbeam Corporation.[1] It was a friendly acquisition of longtime partners that had for more than twenty-five years cooperatively manufactured the *Coldspot* refrigerators for Sears. On the occasion of the union, the new company's work force swelled to more than 5,000 employees.

The refrigerator seen in this photo was the first household refrigerator that the Seeger Company produced as part of the company's conversion from wartime manufacturing to peacetime production. The photo was taken on August 9, 1945. Standing left to right are Seeger executives John S. Holl, Vice President and General Manager of the St. Paul Division; Robert M. Reay, Manager of Commercial Sales; and Neil H. Griebenow, Vice President and Works Manager; kneeling left is Kern H. Sosey, Superintendent; and right J.J. Leonard, Vice President and Sales Manager. Photo courtesy of the St. Paul Pioneer Press.

The new company became a public corporation on March 27, 1946, when the first shares of the Seeger-Sunbeam Company began trading on the New York Stock Exchange. For little more than a year the name of the corporation was Seeger-Sunbeam when in late 1946 the name reverted to Seeger Refrigerator Company, as it had been known for more than four decades.[2] The combined companies continued to manufacture refrigerators and freezers for Sears, Roebuck and Company, Montgomery Ward, and the Admiral Corporation. In addition it produced commercial refrigerators and display cases for the Fridgidare Division of General Motors and various refrigerator cabinets for Westinghouse. In these first years of peacetime, Seeger was manufacturing more than ten percent of the nation's refrigerator production at its plants in St. Paul and Evansville.[3]

Following the end of hostilities with Japan on August 15, 1945, the War Production Board of the federal government immediately canceled all the con-

1945 Annual Report of the Seeger-Sunbeam Corporation

This report covers the operations of Seeger Refrigerator Company, St. Paul, Minnesota, from September 1, 1944, to March 31, 1945, as of which latter date it was merged with Sunbeam Electric Manufacturing Company, Evansville, Indiana, and of the two former companies now known as the Saint Paul and Evansville Divisions of Seeger-Sunbeam Corporation from April 1 to August 31, 1945.

During the past year a major portion of the production facilities of the corporation was devoted to the manufacture of equipment for various war agencies. Sales of war products represented approximately 70 percent of total sales reported therein. With the sudden ending of the war about the middle of August, substantially all of our war contracts were terminated.

A 1945 Seeger-Sunbeam stock certificate for 100 shares. During 1945–46 the corporation was briefly known by this name. Photo courtesy of the Whirlpool Corporation.

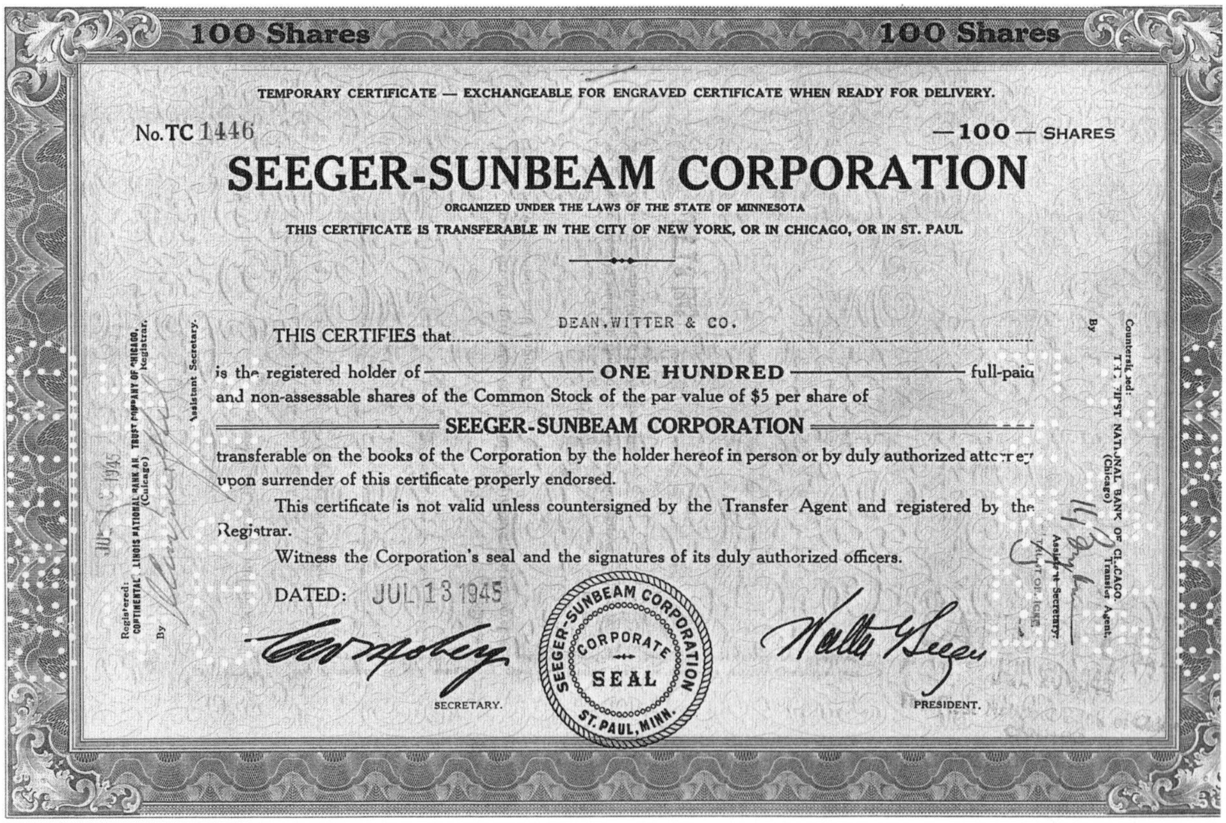

100 Shares 100 Shares

TEMPORARY CERTIFICATE — EXCHANGEABLE FOR ENGRAVED CERTIFICATE WHEN READY FOR DELIVERY.

No. TC 1446 —100— SHARES

SEEGER-SUNBEAM CORPORATION

ORGANIZED UNDER THE LAWS OF THE STATE OF MINNESOTA

THIS CERTIFICATE IS TRANSFERABLE IN THE CITY OF NEW YORK, OR IN CHICAGO, OR IN ST. PAUL

THIS CERTIFIES that DEAN, WITTER & CO.

is the registered holder of ONE HUNDRED full-paid and non-assessable shares of the Common Stock of the par value of $5 per share of SEEGER-SUNBEAM CORPORATION transferable on the books of the Corporation by the holder hereof in person or by duly authorized attorney upon surrender of this certificate properly endorsed.

This certificate is not valid unless countersigned by the Transfer Agent and registered by the Registrar.

Witness the Corporation's seal and the signatures of its duly authorized officers.

DATED: JUL 13 1945

SECRETARY. PRESIDENT.

SEEGER-SUNBEAM CORPORATION CORPORATE SEAL ST. PAUL, MINN.

The Arcade Street plant of the Seeger Corporation in 1950. When this aerial photo was taken, the Seeger facilities covered about sixteen acres of floor space. Photo from the author's collection.

tracts it had with Seeger. After a forty-one month period of wartime production, the company returned on August 9 to manufacturing domestic and commercial refrigerators and a new product, freezers. Predictably, the hardships, rationing, and shortages that had characterized the war years let loose a tremendous consumer demand for household appliances once peace returned. At the same time a building boom gripped the United States during the late 1940s and 1950s in cities and towns across the country because all types of construction had been greatly curtailed during the war. As way to cope with housing shortages that had arisen between 1941 and 1945, many homeowners had taken in borders or let rooms to relatives or friends. Suddenly cities and towns across the country faced a severe housing shortage that was exacerbated by the return of thousands of veterans from overseas. Over time local governments, building contractors, and real estate developers marshaled the resources that were necessary to respond to the demand for housing. The resulting boom in new construction added fuel to the growing demand for home refrigerators and freezers.

An aerial photo of Seeger's main Evansville plant that was taken about 1945. From a Seeger brochure in the author's collection.

Once again events, good timing, good fortune, and keen business leadership were contributed significantly to the Seeger Company's success.

The company's management immediately announced to its employees its plans for a rapid changeover to the production of refrigerators A large number of workers were put to work checking the equipment throughout the plant to put the machinery in first-class operating condition. Tentative plans called for production to start immediately after a brief shutdown to conduct an inventory. Only forty-five days later, the familiar refrigerator cabinets began to trickle from the ends of the assembly lines.

Members of the Purchasing Department rapidly sent orders to all Seeger suppliers to rush the hundreds of various parts such as steel, insulation, gaskets, shelves, screws, bolts, hardware, and numerous other items required for the manufacture of refrigerators and freezers to the company's plants. These orders were critical because before production could start all the necessary parts for Seeger refrigerator cabinets had to be in the plant. Consequently the Purchasing Department had to double its personnel and work overtime to get these materials into the plant. While the company waited for all these supplies to begin flowing to the Seeger docks, workers on the plant floor began turning out sample production runs of each size and model of refrigerator and freezer to acquaint supervisors and their crews with the procedures of assembly. As part of this process, machinists and engineers checked the many jigs that would be needed for full production and issued work requests for any necessary redesign of these jigs if it could improve the quality and/or increase the production of the cabinets.

In addition to undertaking the complex task of shifting resources from wartime to peacetime production, Segeer's corporate leaders faced the urgent need to develop a strategy to upgrade the facilities. In the autumn of 1945 Walter Seeger announced plans for the expansion of the St. Paul plant to enable the factory to increase its annual production of refrigerators from the prewar output of 300,000 to more than 400,000. It was an expansion program that required an investment of $1,200,000 in the plant, and included five new buildings and considerable new equipment that would increase the production capacity of the factory by one-third.[4]

A Sears advertisement for its Coldspot Spacemaster *refrigerator from the early 1950s. This piece shows how the* Coldspot *line had changed from its prewar predecessors. From a Sears brochure in the author's collection.*

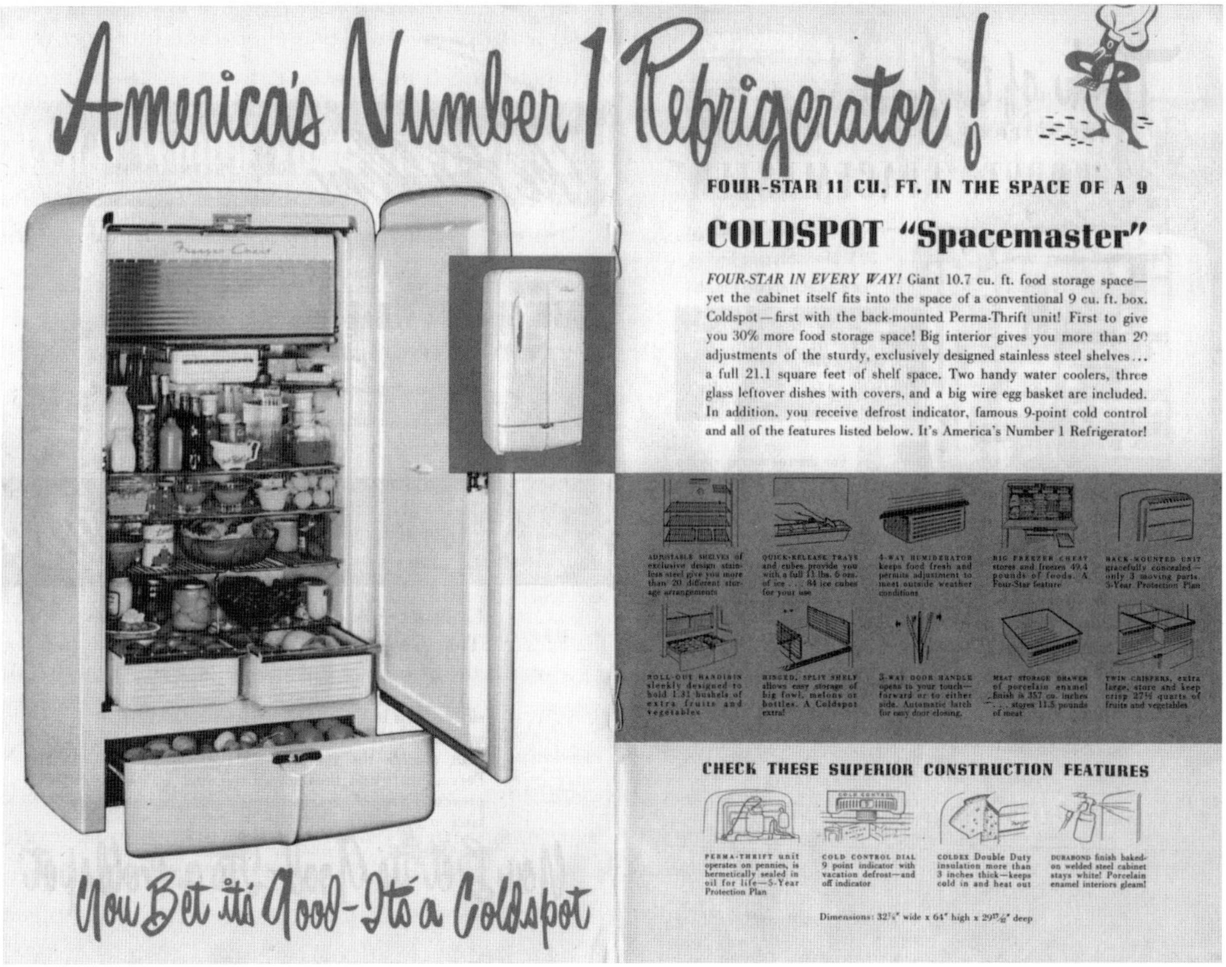

Voices *from the* **Plant Floor** *at* **Seeger**

You Worked for Your Money

"I began work at Seeger on June 17, 1947, at ninety-two cents an hour, first in the steel issue department and then in the metal shop for two or two-and-a-half years, making outframes for refrigerators and freezers. I worked the shift from midnight to 7 A.M. and we had lots of overtime, Saturdays and Sundays during the Korean War. Then I worked on the freezer assembly line, where we manufactured 20 units an hour. You worked for your money. Many persons hired to work in the factory in the morning, left by break-time or the lunch hour. My brother-in-law, Ralph Scalze, was the business agent for the Teamsters Union. My son Steve worked at the factory in the summers during his college years, the income helped to pay for his education. Seeger was known as the old slave place, but I liked the guys. The people worked together, individually or in a group. I had a lot of good bosses. Neil Griebenow carried the axe and always shook hands on inspection days—to see if the machinery was in working order and that the employees were wearing their safety equipment, and was always asking for suggestions for improvement. John Holl walked through the plant and Paul Bowman [did] too. They talked to the men about their jobs."

Orrie Heasley

Walter G. Seeger, president of the Seeger-Sunbeam Corporation at his desk in his paneled office in June 1945. Photo in the author's collection.

John Seeger Holl in 1945. He was then vice president and general manager of the St. Paul plant. Later John Holl succeeded Walter Seeger as president of the family firm. Photo in the author's collection.

Gustave R. Seeger, vice president of manufacturing operations and treasurer of the family firm, in late 1944. Photo in the author's collection.

The improvement and modernization of the physical facilities included a new power plant and wood mill building with new boilers and new 150-foot high stack. East of the Metal Shop building, the company built a steel storage building with overhead, traveling cranes that had the capacity to lift ten tons. The first floor of this new building had a 28-feet ceiling and the third floor housed part of a new commercial assembly line. Two floors were added to the Porcelain Enamel plant. An expanded metal assembly operation was housed on the second floor of this addition and the third floor was used for an enlarged Experimental Department and Laboratory. Part of the new continuous Process and Finishing Department for commercial refrigerator cabinets and display cases was also located on this floor. Some changes in location occurred as well. Thus the entire assembly operations of the Commercial Line of Reach-in and Display Case Cabinets were moved to the third floor of the two new buildings as well as to the third floors of the original buildings. The assembly of the metal case or shell and other operations was also done on this floor. The Metal Shop, where metal parts of the refrigerator cabinet were processed, had some of its equipment relocated. After all the repositioning was completed, the Metal Shop contained the shears, horn presses, gang punches, and large power presses that turned a flat piece of steel into the finished form of the cabinet parts. A new line of large power presses, power brakes, and special equipment to form the sides and top of a refrigerator cabinet from one piece of steel were also located in this department.

The company's expansion plans included the installation of a central cafeteria for employees on the second floor. The building that formerly housed the Storeroom and Express Department was removed, and two buildings, one of four floors and the other of five floors, were erected. The top floors of these two buildings connected with the Domestic Assembly Floor and were used for the assembly of freezers and expansion of the household assembly line. A Test Room was installed for testing the freezer cabinets. A new continuous Black Dulux and Panelyte finishing department was located on the second floor. Equipment for forming, dehydrating, and testing refrigerant coils and tubing was also installed on this floor. The Maintenance and Jig and Die Departments were located on the first floor of this building. The remaining space in these buildings was allocated to stock rooms and replacement parts facilities.

When it was completed, the Seeger expansion project increased the operating floor space of the plant to more than a half million square feet and boosted employment in the plant to 2,300 factory employees. According to an article on the Seeger acquisition of the Sunbeam Corporation in an April 1947 issue of *Investor's Reader,* average hourly wages at Seeger had increased from 66¢ per hour (the equivalent of $8.66 in 2005) in 1939 to $1.14 per hour ($10.79 in 2005) in 1947, which was a substantial hourly wage in those initial postwar years.

Seeger also began the postwar era by launching in mass production of a new appliance—the freezer.[5] As early as 1939 the company had produced a limited number of freezers. This product was actually a revamped milk cooler. In retrospect, this freezer laid the groundwork for Sears and Seeger to become pioneers in the manufacturing and marketing of freezers. Shortly after this product was placed on the market, the top managers of the two companies saw the possibilities of a great future in this new appliance field.

During the decade following the close of

Voices *from the* **Plant Floor** *at* **Seeger**

Seeger Was Like a Big Family

"I got a job at Seeger in 1947, working with the electrician on the night shift at forty-eight cents an hour. My job was to change the tips on the spot welders and was it cold in the factory during the winter months. For thirty years I worked in the pressroom making parts and for twelve or fifteen years I worked in the salvage department, bailing what came to the back door and sending the steel to Malleable Iron Company.

"A guy had a lunch counter as you came in the factory door and you could buy a Williams box lunch for thirty-five cents: two sandwiches, a pie, and an orange. He sold milk, coffee, and candy bars too.

"When the place was running, it was like a big family. It was a fair place to work. I worked thirty-eight years at the company and can count on one hand the days I took off."

John Diesslin

Voices *from the* **Plant Floor** *at* **Seeger**

The St. Paul Saints and Seeger

"I played for the St. Paul Saints baseball team during the 1940s when Walter G. Seeger was president of the club. He arranged for me and several other members of the team to have jobs at Seeger. I started work at the company in 1947, working in the accounting department. The head of the department was Tony Gebhart, who was also treasurer of the Saints. I worked in the general accounting section for about five years and then in cost accounting—a section that crossed all departments of the company. For personnel in management the hours worked were simply as many as necessary. I worked for Seeger, Whirlpool-Seeger, and Whirlpool until my retirement in 1983."

Robert Kelly

Seeger Company officers between 1944 and 1950.

Robert S. Ahrens, chief engineer. He died in 1944. Photo from the author's collection.

Neil H. Griebenow in 1945, vice president and the works manager for the St. Paul and Evansville plants. Photo from the author's collection.

Clarence W. Moberg in 1945. He was secretary and treasurer. Photo from the author's collection.

Irving A. Warren, controller, St. Paul Division. Photo from a Whirlpool-Seeger handbook in the author's collection.

Charles Muessel in 1945, vice president and manager of eastern sales. His office was in New York City. Photo from the author's collection.

Kern Sosey, St. Paul plant supervisor in 1945, succeeded Neil Griebenow as works manager. Photo from a Whirlpool publication in the author's collection

John W. Seeger, vice president and director. Photo courtesy of Lynn Seeger Foster.

World War II, the freezer industry developed rapidly. Sears played a very important part in this growth, recognizing that freezers would play as major a role in the appliance field as the household refrigerator had during the 1930s. Sears spent large sums of money investigating the possibilities of the freezer market. They saturated a small Eastern locality with freezers. Practically every home in the test area had a *Coldspot* freezer. The company undertook extensive surveys to determine the degree to which these freezers provided better living conditions to the families who participated in the tests, the actual savings these families got from having a freezer, and the more appealing lifestyle that was available to these families because of the greater convenience offered by having a *Coldspot* freezer in their homes. After analyzing the market information gained in these and other surveys, Sears and Seeger decided to go all out on freezer production and sales.

Several newspaper employment ads from the late 1940s for positions which were open at the Arcade Street plant. These ads are from the St. Paul Pioneer Press, *the* Seattle Times, *and the* Detroit Free Press. *From the author's collection.*

Because the residential freezer was an entirely new product for Seeger, the company's managers took steps to maintain their high standard of quality along with their established reputation. In part this meant that numerous samples had to be prepared for completion of Engineering, Experimental, and laboratory tests. Seeger had established a reputation in the refrigeration field for quality merchandise and they did not want to jeopardize it. Once these sample freezers passed all the company's quality control

Voices *from the* Plant Floor *at* Seeger

Hired at Seeger in 1949

John Herman, who was physically and mentally challenged, was employed at Seeger for thirty years (1949–1979). His mother died at birth and he was raised by his father at their home on Rice Street. During his early elementary school years he attended in succession the Smith, Gorman, and St. Bernard schools to the third grade but did not learn to read or write. He also was a patient at the Gillette Hospital for Children. As a handicapped person he was denied employment by several St. Paul businesses. He learned that Seeger hired handicapped persons and in 1949 began his career at the Arcade Street plant. His primary job was to wash down the refrigerators, freezers, and milk coolers on the assembly line. During periods of lay-off at the factory, he worked as a janitor rather than lose the income.

Herman recalls being paid weekly at first and later every two weeks, becoming a member of the Credit Union, and buying U.S. Savings Bonds. He remembers such other benefits of employment as health coverage with Aetna and Blue Cross, and a Prudential group life insurance policy. For his long service he received the traditional awards of recognition: for ten years of service a pin with a ruby stone; for twenty years a pin with a diamond; and for twenty-five years a wrist watch. His association with Seeger allowed him to purchase and maintain a house on Blair Street, raise and see a son graduate from college, save money, and to be independent during his retirement years.

John Herman

measurements in terms of engineering standards, design requirements, and production efficiency, they started to roll off the Arcade Street factory assembly lines.

Between 1939 and the start of World War II, Sears had sold a limited quantity of freezers. This initial sales effort had given their marketers some awareness of the potential demand, but the war prevented any further sales efforts until peace returned. Following the receipt of production orders from Sears in 1946, Seeger produced a total of 11,877 freezers for this valued customer. The following year they more than doubled this production with a total of 25,081. Production again doubled in 1948 with approximately 50,000 freezers delivered to Sears and sold under the trade name of *Coldspot*. In 1950, Seeger produced 124,832 freezers and in 1951, 163,208 *Coldspot* freezers.

Carl Bjornerantz, the distinguished industrial designer, redesigned the 1951 Sears line of home freezers with the aid of the Seeger engineering department. He based the new design on two special features: the use of high-density insulation that made possible increased interior capacity and decreased exterior dimensions, and new production equipment and methods that resulted in manufacturing economies. The new Sears freezers won national design acclaim. Steel strikes in 1952, however, caused shortages in the manufacture of the component parts for freezers, which set production back. Consequently freezer output at the Seeger plant during that year decreased to 123,266 units, almost a twenty-five percent drop in volume from 1951.

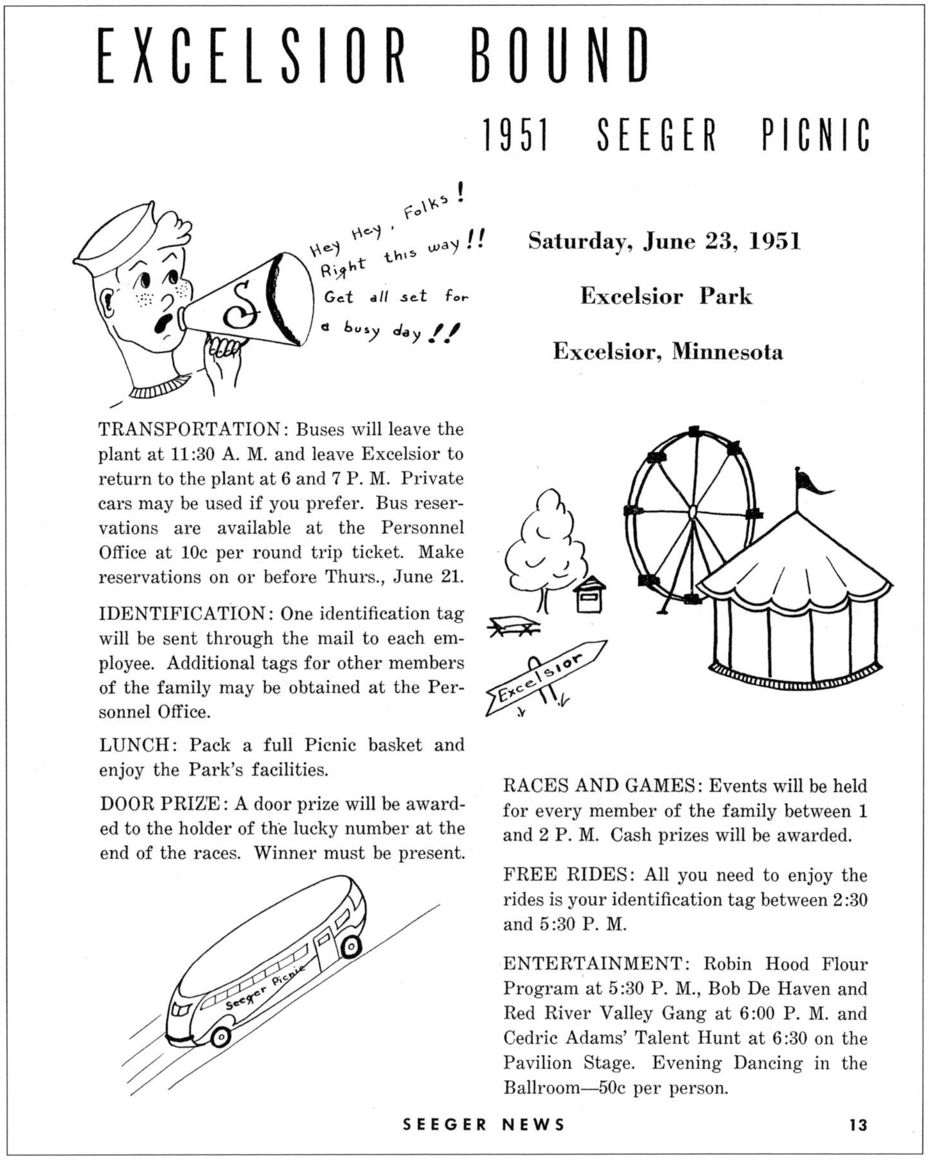

EXCELSIOR BOUND

1951 SEEGER PICNIC

Saturday, June 23, 1951

Excelsior Park

Excelsior, Minnesota

TRANSPORTATION: Buses will leave the plant at 11:30 A. M. and leave Excelsior to return to the plant at 6 and 7 P. M. Private cars may be used if you prefer. Bus reservations are available at the Personnel Office at 10c per round trip ticket. Make reservations on or before Thurs., June 21.

IDENTIFICATION: One identification tag will be sent through the mail to each employee. Additional tags for other members of the family may be obtained at the Personnel Office.

LUNCH: Pack a full Picnic basket and enjoy the Park's facilities.

DOOR PRIZE: A door prize will be awarded to the holder of the lucky number at the end of the races. Winner must be present.

RACES AND GAMES: Events will be held for every member of the family between 1 and 2 P. M. Cash prizes will be awarded.

FREE RIDES: All you need to enjoy the rides is your identification tag between 2:30 and 5:30 P. M.

ENTERTAINMENT: Robin Hood Flour Program at 5:30 P. M., Bob De Haven and Red River Valley Gang at 6:00 P. M. and Cedric Adams' Talent Hunt at 6:30 on the Pavilion Stage. Evening Dancing in the Ballroom—50c per person.

SEEGER NEWS 13

Ad announcement from the Seeger News *informing employees of the 1951 Seeger Company picnic at Excelsior Park. From a* Seeger News *in the author's collection.*

The Freezer Assembly Line

What was the process on the Seeger assembly line? Many hours of hard and thorough work were required before finished freezer parts headed for the freezer assembly line. This line was the final step in the manufacture of the *Coldspot* Home Freezer. In the summer of 1954 there were two assembly lines operating on the day shift and the 170 workers on these lines assembled 520 freezers a day.

Briefly, the assembly line process was launched when the freezer shell arrived on the line from the Dulux paint shop where it had been thoroughly cleaned, sanded, and given two coats of DuPont enamel. An ultramodern

Voices *from the* Plant Floor *at* Seeger

Looking for Scratches and Dents

"My first day on the job at Seeger was December 11, 1951, and I packed my lunch. I came from Laredo, Texas, and my first job was to put tar in the liners of freezers. After one month I worked at a machine press for the next ten years. Later I became an inspector after products came off the assembly line—looking for scratches and dents. I was a group leader of other inspectors.

"I was on a Seeger bowling team from 1958 to 1984. We were champions in 1972, 1973, 1980, and 1981. My [Seeger] badge number at the beginning was 1703; a week later it was 1702. At the close of the plant in 1984 it was 299. I missed only three days of work in thirty-four years and received two watches and a service pin."

Raol Gomez

sprayer system that was installed in the plant in 1949 painted these shells. Upon reaching the line, the shell was mounted on a skid, placed on the assembly line, and started on the last lap of its journey. Line operations were divided into five categories: Set-up Group, Moulding Group, Unit Group, Door Group, and Crating Group. The Set-up Group operations began with the bolting of a skid that facilitated handling of the freezer to the bottom of the shell. When this had been done, the shell coils received a coat of Thermalmastic that aided in conducting heat. Then the shell was placed on the line, sealed with hot asphalt, and insulated with seventeen pieces of fiberglass. Seventeen pieces of tightly packed fiberglass insulation were used in each freezer cabinet for bottom and sides. At this point the shell was ready to receive a liner. A belt carried the liners from the floor below in the factory to within a few feet of the installation spot. While the liner was being inserted in the shell, guides were attached to the shell to prevent damage to the insulation. Liners arrived at the point of installation directly from the coil room. After the liner was fitted, Masonite breaker strips were inserted. These strips did two jobs: they sealed the top of the cabinet and fastened the liner to the shell.

Spot-welding the hinges on a freezer in the Arcade Street plant in 1953. From a Seeger News *in the author's collection.*

The cabinet then moved into the Moulding Group where a vapor-leak test was made. The vacuum tester was an exclusive development of Seeger's and had been in operation since 1949. Each cabinet assembled had to hold an even pressure on the tester dial for twenty-eight seconds. If a cabinet could not hold pressure, it was inspected for leaks that were sealed with either Permagum or a mastic sealer. Each cabinet went through several tests so that it would be in tip-top working order. A Freon leak test checked to make certain that no

An assembly line worker positions a freezer shell on the conveyor that went to the Dulux paint shop. From a Seeger News *in the author's collection.*

refrigerant was escaping; a vapor seal test made certain that the cabinet did not leak; and a Diaelectric test satisfied inspectors that there were no electrical shorts in the units.

The Laboratory maintained a test room on the freezer production line for the purpose of checking the quality of performance of the freezers. This insured that the cabinets produced would not only be satisfactory in appearance and construction, but would also operate as efficiently as they were designed to do.

The test room consisted of an insulated, air conditioned, space maintained at a temperature of 80° F, that was equipped with the necessary recording and indicating instruments that enabled the operator to determine whether the freezer maintained the required temperatures without excessive or inefficient operation. For each test required the use of a two-pen, recording pressure gauge, three temperature-measuring devices, two electric clocks, and one kilowatt-hour meter. The pressure recorders were individually spring-mounted at each test station so that connection to the condensing units could easily be made. The remaining instruments were mounted on a test panel at which the operator could take periodic readings. An analysis of the data taken indicated the temperatures maintained through the cabinet, the percentage of the time the freezer had operated and the frequency of operation, the power consumed by the motor, and the refrigerant pressures at which the unit had operated. Before the cabinet under test was again released to the production line, a recheck was made for electrical leakage from the motor wiring, refrigerant leakage from the refrigerating system, and another check that the motor overload device was operating properly. The final part of the test was to remove a sample of the refrigerant for a moisture analysis by the Chemical Laboratory.

Two freezers were selected at random from the production of each shift, moved into the test room, and tested for a sixteen-hour period after the cabinet temperature had reached. If a defect of any sort was found, the appropriate production and inspection personnel were immediately notified so that efforts could be intensified to prevent recurrence of this defect. Weekly reports on these test results kept key departments aware of any quality issues and careful analysis of these records enabled the Engineering Department and the Laboratory to predict future performance trends and assure customers that Seegers freezers met all company quality standards.

After testing, the freezer moved into the Unit Group where a rubber collar was fitted around the top and the cabinet traveled through an infrared oven. It was baked at 180° F, which evaporated any residue from solvents and prevented any unpleasant freezer odors. Spot testing was done to insure the effectiveness of the baking process. After the collar was in place, a unit was installed in each cabinet. Units stood at a constant temperature in the Unit Soaking Room for four hours to make certain that they carried a proper refrigerant charge.

Workers complete the final metal finishing touches on a freezer shell prior to its arrival in the Dulux paint shop. From a Seeger News *in the author's collection.*

The Door Line was a separate operation. Doors were lifted from a conveyor, insulated, and completed with an inside panel, handle, and plastic gasket. When the door was finished, it was hung on the cabinet. When the door was hung and a test run made, the freezer was complete and ready for final inspection. The primary job of the Inspection Department was to make a final check on the general appearance of each box, see that the doors sealed properly, and to ensure that the liners were in the correct position in the cabinet. A preliminary inspection, or trial run, was made and any defects were corrected. After inspection the freezer started on its trip under Arcade Street to the Crating Group in Building 17 on the west side of the street. When it arrived,

Finished freezers prior to crating and shipment to customers. From a Seeger News *in the author's collection.*

freezer compartment dividers and storage baskets were placed inside and the freezer was securely packed in a shipping crate. The *Coldspot* Home Freezer was now ready for delivery, via railroad or truck, to Sears, Roebuck stores and warehouses throughout the United States.

Because Sears had pioneered in developing the market for freezers, the rest of the freezer industry saw the light and began to increase the promotion of home and farm freezer sales. The market was growing, and Sears believed that the more the entire industry boosted sales, the greater the possibility that Sears would improve its sales. In other words, freezers actually advertised and sold themselves through proud and pleased owners. By 1953 the freezer market was saturated at approximately twelve percent. That means that twelve families out of every one hundred owned a freezer, a factor that suggested a vast potential market for future freezer sales.

Sears and Seeger developed a new line of freezers for 1953 and those models began coming off the Arcade Street plant's production line in December 1952. By adding color and other features to the line, Sears and Seeger again led the industry. Seeger also added a new line of appliances that year, a vertical or upright-type freezer. Detailed public surveys indicated that there was a definite demand for an upright or vertical-type freezer model, especially among consumers who lacked the floor space for a horizontal or chest-type freezer. To increase freezer sales, Sears introduced two upright freezer models. One was a standard model that had the same operational features as the deluxe line but did not have all the design or appearance features; the other was the deluxe model.

Across the United States, Sears conducted a series of promotional "Caravan Sales Meetings" for the purpose of showing their new lines of appliance merchandise to their store managers and sales personnel. The gatherings were held in fifteen key cities and every meeting generated an enormous amount of enthusiasm for the Sears products. Following this rollout of the new freezer lines, Sears sales personnel set their 1953 sales quota for freezers at $100,000,000.

What Makes a Freezer Freeze?

Because the Seeger Engineering, Research, and Experimental Departments were intimately involved in the technological development of the mass-market home freezer we should raise the question: What makes a freezer freeze? Mechanical refrigeration as used in all parts of the world today is basically the same. There are large refrigeration systems with special devices and accessories to make operations safer and more efficient, or small refrigeration systems, such as were built in the Seeger plant, with only those parts necessary for the simplest and safest operation. Even the smaller and simpler mechanical systems had only five basic parts. These parts were the compressor, the condenser, a restrictor or expansion device, an evaporator, and the all-important refrigerant.

Starting at the intake of the compressor, refrigerant in the form of a vapor is drawn in and compressed to a high pressure of about 190 pounds pressure where the room temperature is 80° F. As the refrigerant is compressed, it becomes warmer for the same reason a hand-operated tire pump gets hot when air is compressed in it. This hot, high-pressure vapor then leaves the compressor and enters the condenser.

In the condenser the vapor is cooled and changed into a liquid, much as cooled steam changes back to water. As the refrigerant leaves the condenser and enters the capillary tube, it has cooled to slightly above room temperature. As the refrigerant flows through the capillary, the pressure on it becomes lower. As the pressure drops, so does the boiling point of the refrigerant. When the refrigerant reaches the outlet end of the capillary, it has dropped to a pressure of ten pounds per square inch, but the temperature of this refrigerant, such as Freon 22, one of several refrigerants used by Seeger, has dropped correspondingly to a temperature of -20° F.

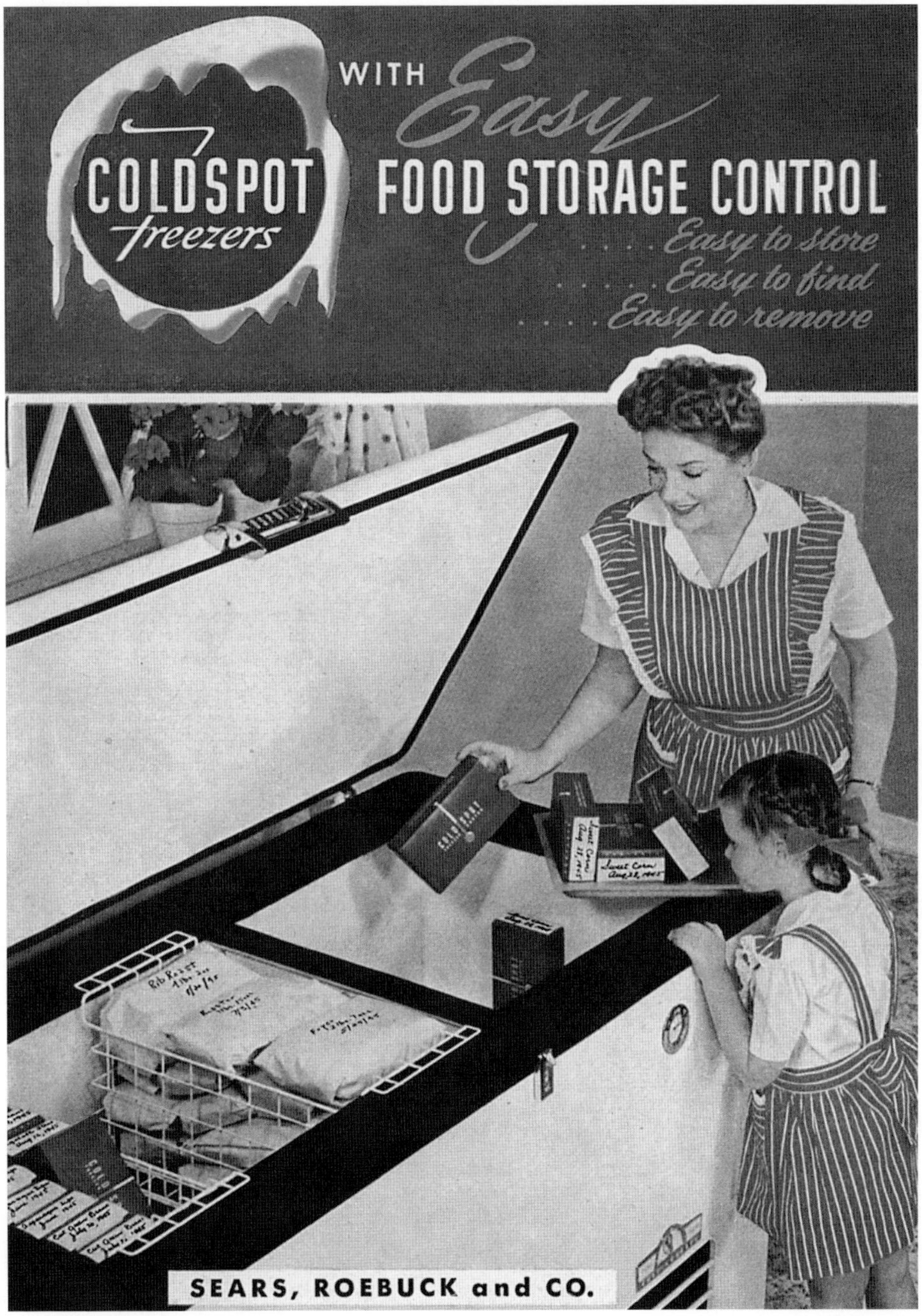

A 1950s Sears advertising brochure for its Coldspot *freezers. From a Sears brochure in the author's collection.*

Next this boiling refrigerant, still at -20° F, reaches the evaporator or cooling coils and begins to cool the interior of the freezer. As the refrigerant boils and cools, it turns into vapor, the same as boiling water turns into steam. This vapor passes back to the intake of the compressor and is again compressed as described at the beginning of the process. During its passage through this cycle, the refrigerant has absorbed heat from the interior of the freezer and carried it to the compressor.

Seeger engineers used two different types of compressors in freezers built at the plant. The rotary-type compressor, built at the Evansville, Indiana, factory, was used in the 4-cubic-foot and the 10-cubic-foot chest-type freezer. Both the 14-cubic-foot and 20-cubic-foot chest-type freezers and the 18-cubic-foot upright freezer used Tecumseh reciprocating or piston-type compressors.

The function of the condenser tubing on a freezer was to discharge the heat drawn from inside the freezer. Small amounts of heat constantly leak into the cold space of a freezer through the insulation, just as heat has a tendency to go out of a building through the walls in the wintertime. It was this heat and also the heat in warm food placed in the freezer that had to be removed.

On the 10- , 14- , and 20-cubic-foot chest-type freezers, and on the 18-cubic-foot upright, this condenser was made of steel tubing fastened to the inside surface of the exterior shell. To increase the contact of the tubing to the shell, Seeger used a mastic that was a good conductor of heat. When hot refrigerant vapor left the unit and entered the condenser, it was cooled down by this tubing, mastic, and shell combination. As it passed further along, the refrigerant began to condense or change to a liquid.

On the 4-cubic-foot freezer, steel tubing was again used for a condenser. In this case it was a group of coils in the air in back of the freezer. These coils had fins on them to increase the surface exposed to the room air. This method of discharging heat was formerly used on all freezers.

Seeger used capillary tubes as restrictors or expansion devices on all its freezers. Their purpose was to maintain a pressure difference between the condenser and evaporator, and still let the refrigerant flow through at a prescribed rate.

High pressures were maintained in the condenser while lower pressures were maintained in the evaporator. The compressor increased the pressure in the condenser and provided the flow through the system. The capillary had exactly the opposite effect. It let the pressure drop down again between the condenser and evaporator, but it still accommodated the flow of refrigerant. Capillaries were made of copper tubing with an inside diameter about the size of the wire in a paper clip. They varied in length from 5 feet to 18 feet, depending on the size of the freezer.

On all chest type freezers, freezer evaporators or cooling coils consisted of copper tubing wrapped around the porcelain freezer liner. The 18-cubic-foot upright had coils of aluminum tubing welded to aluminum shelves. On the chest-type, Seeger used mastic that was a good conductor of heat to insure contact between the tubing and the liner. The liquid entering the evaporator tubing drops in temperature, as the pressure drops, to the above-mentioned -20° F. The refrigerant in the evaporator tubing absorbs the heat from the food and any heat that has leaked in through the insulation before the refrigerant returns to the compressor intake.

In order to increase the efficiency of the system, a heat exchanger was added. As the refrigerant leaves the end of the evaporator coils, it is still cold. The liquid refrigerant coming in through the capillary is warm. Rather than let this cold go into the room, which is what would happen if something were not done, the tubing leaving the freezer was soldered to the capillary coming in. In this manner, the cold vapor leaving the freezer is warmed and the warm liquid entering the freezer is cooled through an exchange of heat. This exchange saved some of the refrigerant effect available in the liquid refrigerant leaving the evaporator.

Seeger engineers used Freon 12 as the refrigerant in the 4- and 10-cubic-foot freezers (the chemists call Freon 12 dichlorodifluoromethane). Freon 22 was used in the 14- and 20-cubic-foot chest-type freezers and the 18-cubic-foot upright freezer (chemist's name for Freon 22 is monochlorodifluoromethane). The property that made both kinds of Freon so valuable was the fact that they boiled at such extremely low temperatures. These refrigerants are, when properly used, non-inflammable, non-toxic, and non-corrosive. Freon is dangerous when in contact with an open flame; the result is phosgene, a poisonous gas of a type used in warfare. Freon, when in contact with small quantities of water, forms an acid that will over time corrode parts and cause trouble. Consequently great care was taken to remove all moisture during manufacture.

Other important parts of the freezer, but not necessary to the refrigeration system, are the drier, the accumulator, the control mechanism, and the insulation. Freezer assembly at Seeger included a drier that contained two cubic inches of silica gel. It was located between the outlet of the capillary tube and the evaporator. Its purpose was just what its name implies. Silica gel would attract and hold moisture to the extent of removing enough to be below the safe level if proper care was taken when building the freezer.

Located at the outlet of the evaporator coil was the accumulator, the purpose of which was to act as a shock absorber when the unit starts. Its action was to hold a large quantity of liquid refrigerant when the compressor started. If it were not present, liquid refrigerant could go right back to the compressor and under certain conditions cause harm.

All Seeger freezers used Fiberglas insulation. It was 3½″ thick in the bottom and 3″ thick in the sides. The lid had insulation that is 2½″ thick.

A thermostatic control regulated the temperature of the freezer. It was

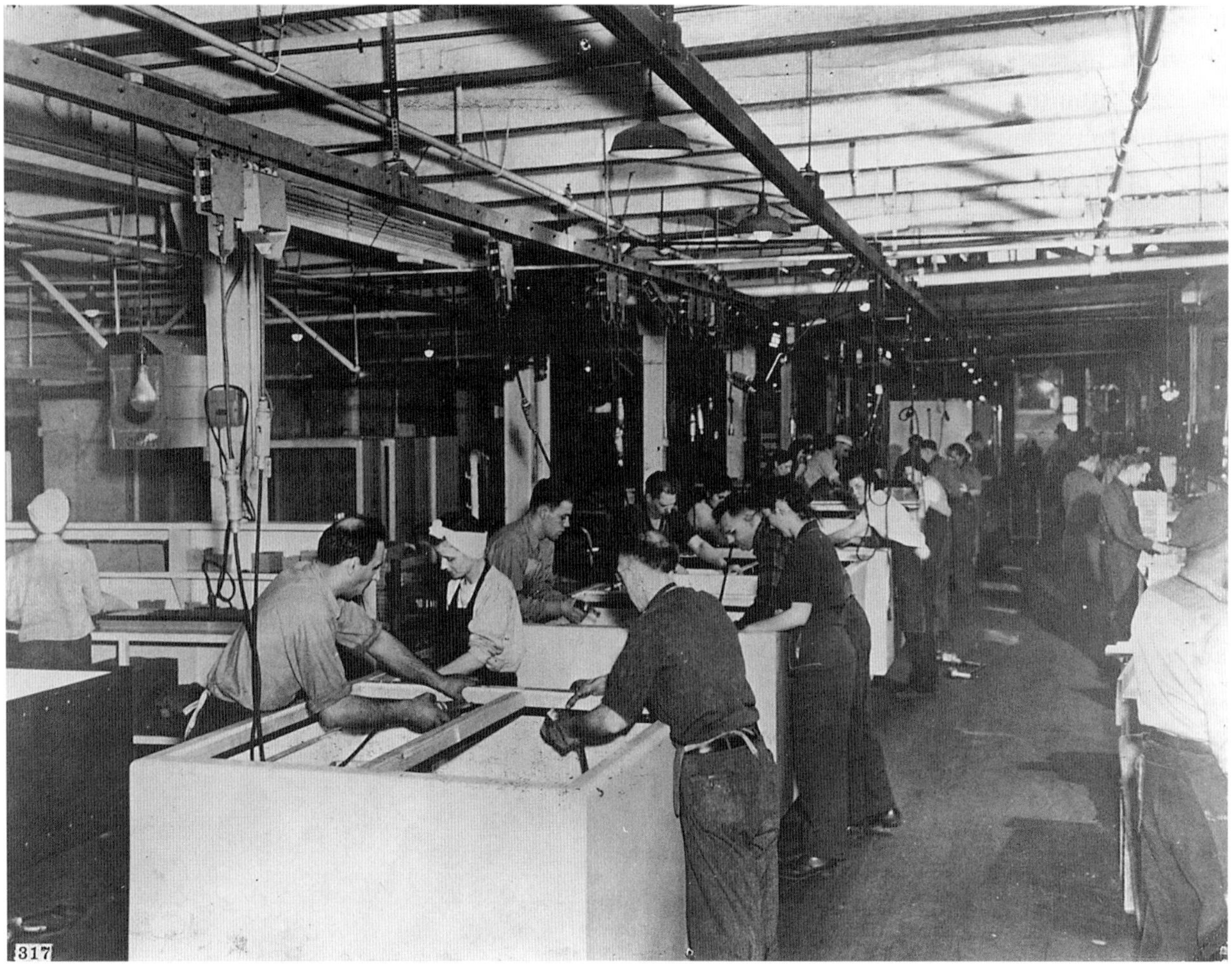

The Seeger freezer assembly line in 1948. Photo courtesy of the Minnesota Historical Society.

located and adjusted so the freezer would cycle to hold the temperature virtually constant. The control was designed so that failure would generally cause the freezer to operate all the time rather than shut off and warm up.

Research Department

If Seeger had always restricted its activities to the manufacture of iceboxes, it could never have reached its industrial stature. A cautious side-glance at the competition was stimulating and informative, but it would be of no use if the company lacked the technical know-how. Beginning at the turn of the twentieth century, a few of the more enterprising industries in the United States began to see the value of technology as applied to their particular fields of endeavor. The basic idea spread rapidly, gaining momentum like a snowball rolling down hill and developing to the point where technology quickly changed from a luxury to an economic necessity. For some it was a bold forward step that paid off at the cash register, for others a defensive gesture necessary to assure their future competitive position in the business. Warfare in the first half of the twentieth century placed renewed emphasis on science and demonstrated the value of research and development, first in the area of weapons development and then after 1945 in all sorts of industrial fields. Consequently many businesses intensified their research and development activities to the point where a company's dollar volume of business was frequently in direct proportion to the size of its technical staff. Modern industrial science has given the lie to the age-old fable of the tortoise and the hare, for in a purely technical sense, the race is always to the swift and the battle always to the strong. This is the research point of view; yet as long as such thinking is confined within a laboratory, industrial growth will be disappointingly inadequate because pure ideas rarely make money on their own. Only when skilled leaders implement these new insights or procedures with the necessary financial and material support will they yield a return on the time and effort invested in them. Serious research and development work requires teamwork of a high order was and it can be obtained in only such proportion as the research spirit pervades a company. Unless everyone connected with the implementation of a new idea is truly convinced of its inherent worth, it is apt to wither at some point along the product development line. Thus in the decade after World War II, Seeger managers recognized that the mere physical acquisition of research and development facilities were not in themselves a short cut to increased profits, although admittedly they were a decisive step in the right direction.

A second phase of laboratory operation involved development. This generally involved investigations into the relative merits of new processes required to fit the constantly changing needs, or the improvement of existing processes, or the substitution of new materials in old processes; or, in some cases, it involved the discovery and application of an entirely new idea. The primary purpose of research and development at Seeger was not generally fundamental research. True research ordinarily involved the discovery of previously unknown facts about a substance without regard to their commercial significance; whereas, development took these facts and applied them to the fund of current knowledge, usually impelled by a basic profit motive.

In 1951 Seeger established a new modern research laboratory in the Minnesota Box Building located behind the Arcade Street plant on Forest Street. As the postwar selling boom of home appliances began to subside and an abundant supply of these products was built up in the warehouses of nearly every manufacturer in the country, the retail sales organizations found themselves engaged in stiff competition for the buyer's dollars. To

Labor-Management Relations at Seeger

The Seeger Company's extensive corporate records that had been carefully organized and maintained for the period between 1902 and 1955 were massively and thoughtlessly destroyed at the closing of the St. Paul factory in December 1984. This action had catastrophic consequences for the historian and reader because our understanding and knowledge of the company's labor history over several decades is severely limited. For example, it is impossible to identify the events, years, and employee leaders who initiated the effort to organize the unions at the Seeger factory. Did labor organizing occur initially in the 1920s or '30s? Did unionization take place throughout the factory at one time or was it introduced over a period of years department by department or by occupational specialities such as electricians, millwrights, toolmakers, or sheet metal workers? With the loss of the company's personnel records, there is no way to know with precision the number of employees who worked at the plant or were members of unions at any time during the Seeger history. Detailed information regarding the labor controversies that led to strikes at the factory in 1940 and 1951 has also been lost.

Based of the anecdotal evidence provided by a band of surviving Seeger employees from the 1930s, the company was probably a closed union shop by the end of that decade. The fragmentary records that have survived and information gathered from interviews with former employees indicates that in all likelihood all production employees were required to be members of the Refrigerator Workers Union. The International Union of Electrical, Radio and Machine Workers Union probably represented workers with certain trades and skills. By 1946 about 1,800 manufacturing employees were members of the Refrigerator Workers Union, an affiliate initially of the American Federation of Labor that later became associated in the early 1950s with the Teamsters Union. Several other unions also represented Seeger employees at the time including the International Brotherhood of Electrical Workers, Local Union Number 110; the Millwrights and Machinery Erectors, Local Union Number 548; the International Guards Union of America; and the International Union of Operating Engineers, Local Unions 36, 36-A, and 36-B.

The limited understanding of union operations at the St. Paul and Evansville factories that we have today is due entirely to the records of James Weldon Click, an officer of the Union of Electrical, Radio and Machine Workers affiliated with the Congress of Industrial Organizations and based in St. Louis.[1] The earliest document in the collection, a contract for employees represented by the union at the Evansville plant in 1947, describes the members' rights and benefits including such matters as union security, responsibilities of the company, grievance procedure, terms of seniority, layoffs, discharges and disciplinary layoffs, military service, maternity leaves, and that men and women performing the same job were to be paid the same wage. The company was required to furnish the union with job descriptions showing the general requirements of each job classification. The job descriptions were to be used in determining the proper wage classification. The guaranteed minimum wage after four weeks of employment was set at $1.00 per hour. The vacation period at the factory in 1947 was from August 17 to August 31, a procedure that was followed at least until the merger in 1955 with Whirlpool and RCA. Deliberately reducing the workforce at a manufacturing plant for a certain period of time, such as two weeks in the summer each year, and giving most employees some vacation time was a common practice in many industries in the middle of the twentieth century. Union members with one year's seniority and who had worked 700 hours (essentially 14 hours per week for 50 weeks, which therefore would have included some employees who may have been considered part-time workers) in the preceding year were entitled to one week's vacation with pay while persons with more than five years' seniority were granted two weeks vacation with pay.[2]

The labor contract noted that individual union members at the completion of one years' service with Seeger received life insurance coverage of $500 under a group insurance policy, which was increased $100 per year thereafter until it reached a maximum of $1,000. In 1947 there was no health insurance program for employees, but the company was required to install safety devices wherever needed and furnish soap, towels, gloves, and incidental articles where necessary for the health and safety of employees. The contract also stipulated that the company was required to provide first aid and medical services to any employee who was injured on the job.[3]

The 1950 contract between the union and the Seeger factory in Evansville provided the same rights and benefits as described in the 1947 agreement.[4] In addition the company agreed to deduct from each employee's wages the union initiation fee and the monthly dues for each member of the bargaining unit. The minimum hourly wage after eight weeks' employment was raised from $1.00 to $1.16 and hourly pay rates were set for the numerous job descriptions and classifications at the plant. Seeger was to provide the Standard Blue Cross and Blue Shield hospital and surgical benefits then in effect at the plant. The union member was able to purchase through payroll deductions medical coverage for dependents. Seeger was required to establish a retirement benefit program that would provide $100 per month, including Social Security, at the age of 65 for all employees with twenty-five years' or more continuous service with the company.[5]

An Apprenticeship Program was introduced in the 1947 contract with the union for toolmakers and expanded in 1950 to include electricians, carpenters, millwrights, and sheet metal workers.[6]

In 1951 the St. Paul and Evansville factories of Seeger were hit by a strike of the production and maintenance employees represented by the Refrigerator Workers' Union, affiliated with the American Federation of Labor.[7] Workers excluded from this bargaining unit were electricians, millwrights, truck drivers, engineers, salesmen, guards, electrical employees, and supervisors. The agreement that settled the strike for the St. Paul plant was identical with the rights and benefits granted in 1950 Evansville contract with the International Union of Electrical, Radio and Machine Workers. A new feature was the introduction at the factory of a relief (or break) period of ten minutes that was allowed at the end of each two-hour work period. In addition the Refrigerator Worker Union contract addressed such new issues as the differential pay for persons who worked the night shift, payment of the difference between hourly wage and jury pay when a member were called for jury duty, the minimum wage was increased from $1.16 to $1.33 per hour, and hourly wages were determined by job descriptions and classifications within each department after negotiations between the company and the union.[8]

The 1952 Agreement between the International Union of Electrical, Radio and Machine Workers at the Evansville factory more precisely refined and defined the rights and benefits of members over the 1950 contract.[9] It kept in force the terms of the previous agreement, but the minimum wage after eight week's of employment was increased from $1.33 to $1.41 per hour. Hourly wage rates were established by the job descriptions and classifications for each department.

While acknowledging that those Seeger documents that have survived provide only a fragmentary understanding of labor-management relations at the company, the records that are still available indicate that federal regulations that applied to labor negotiations in the postwar era did influence contract language and terms at Seeger. In addition the gradual introduction of more employee benefits via union contracts paralleled some labor negotiations in other manufacturing sectors such as automobiles. The growing consumer demand during this time for the wide variety of products that the Seeger Company made in conjunction with the economy of scale that was already possible and the increases in productivity that were occurring in the immediate postwar period gave Seeger management cause for steadily increasing its minimum hourly wage and expanding the benefits the company provided to its union workers.

Notes

1. James Weldon Click Papers, Western Historical Manuscript Collection, Thomas Jefferson Library, University of Missouri–St. Louis, St. Louis, Mo.
2. Ibid., box 6, no. 44, 1947 Agreement between Seeger and UERMWA, Local 813.
3. Ibid. For additional details regarding labor relations at Seeger's Evansville, Indiana, factory between 1947 and 1950, see the William Sentner Papers, Washington University Library, St. Louis, series 1, box 1, folder 4, District 8, Local 813, Seeger Refrigerator Company, Evansville, Agreements between Seeger and UERMWA, Local 813.
4. James Weldon Click Papers, box 7, no. 57, 1950 Agreement between Seeger and the International Union of Electrical, Radio and Machine Workers, Local 808.
5. Ibid.
6. Ibid.
7. Ibid., box 7, no. 62, 1952 Agreement between Seeger and the Refrigerator Workers' Union, Local 20459.
8. Ibid.
9. Ibid., box 7, no. 65, 1952 Agreement between Seeger and IUERMWA, Local 808; box 8, no. 83, Seeger Wage Rates, 1950s; and box 20, no. 232, *Seeger Refrigerator Company, Retirement and Disability Pension Plan for Hourly Rated Employees.*

assure itself of a representative share of sales, Seeger needed to produce a product that would strongly appeal to a large percentage of the buying public.

Generally, a person considering the purchase of a home appliance is interested in the appearance, operation, dependability, and cost of the item. Not all of us would agree entirely as to their respective relative importance in determining a purchase, but we would probably all agree that we considered each of these items were important factors in arriving at a decision to buy a particular home appliance. Therefore, Seeger was constantly seeking to improve its products and worked toward cost reductions so that it would continue to have a favorable competitive position in the national market. The Research Department was a vital member of the team that was constantly working in this direction. The department worked closely with both Engineering and Production to perform two principal functions along with a long list of more, specialized services. The first of these entailed the testing of production models as they were produced on the assembly line, and models that were altered in some manner for a number of reasons. A change may have been requested by one of the large buyers, or it may have been a change that was dictated by necessity such as government restrictions on the use of copper. This work was essential to the day-to-day operation of the plant if the high standard of quality was to be maintained and improved.

The second function was to do long-range research and development work on new models or products that may have required from six months to several years to complete. The new laboratory was built to provide additional space, facilities, and personnel to expand the capacity of the Research Department to perform this work. Located on the west side of the Minnesota Box warehouse building, the new laboratory was completely air-conditioned. The equipment located in these rooms was planned so that the laboratory was independent of the facilities in the main plant.

The problems that required investigation and research were not all of a mechanical nature. A chemistry section was set up to solve those problems that dealt with materials used in the fabrication of Seeger products. In general, the primary function of that section was to evaluate new materials and techniques in refrigeration and allied fields. Careful experimentation preceded any evaluations. When enough information was obtained, then the research team would make its conclusions. The actual time spent on a project varied from a few days to a year or more depending upon the ultimate possibilities inherent in the research project.

If the Seeger Company wanted to continue as a leader in the field, it had to be constantly aware of changing consumer tastes as regarded cabinet design features, and accessories, vigilant in its surveillance of the developments of other manufacturers, and willing to maintain a progressive program of investigation and research. The responsibility for these operations was placed on the Sales, Engineering, Experimental, and Research Departments whose job it was to investigate and develop the products that had the greatest sales appeal and contributed to the large production necessary for the welfare of the company. The new Research Laboratory, by increasing the facilities available to conduct its share of the work, provided the organization greater security in maintaining its favorable competitive position and representative share of sales.

Bulk Milk Coolers

In 1955 the Seeger Commercial Assembly Line acquired a new look with the introduction of a new product.[6] In a brief period multiple departments, including Sales, Engineering, Purchasing, Industrial Engineering, the Jig Shop, Maintenance, and production supervision all had a hand in a rapid changeover that involved removing walls, conveyors, equipment, and other material while lighting, power outlets, and piping were rearranged and new and completely different equipment was moved in and installed.. The new look was simply the transformation of an area previously used for the production of commercial refrigerators to an area for the manufacture of bulk milk coolers.

Here a welder joins the ends of the exterior shell of a bulk milk cooler. A custom-designed jig holds the pieces in place for accurate spot-welding. From a Seeger News *in the author's collection.*

Although the visible change took place in a short time, months of planning and preparation had been necessary to make it possible. Experimental models had been built, blueprints had been prepared, and a variety of materials were purchased. The rearrangement of production facilities took place to accommodate existing commercial appliances as well as the new addition.

The Sunset Equipment Company, which had been users of Seeger commercial products for twenty-five years, contracted with Seeger for the St. Paul company to build these bulk milk tanks. The Seeger Company relationship with Sunset Equipment was very successful and it continued for another twenty-seven years after the introduction of the new tanks. Quality was the watchword of the dairy industry. Just as quality control was the constant concern of the dairyman, so was quality in design, materials, and workmanship the primary objective of the Sunset Bulk Milk Cooler. Built of all-welded, nickel-bearing, stainless steel inside and out, and from lid to leg, with evaporator of noncorrosive metal and spun glass insulation, the lifetime bulk milk cooler had at last been created.

The agitator in the Sunset Cooler was a one-piece, slow speed, quick-coupled connection. Its special-type agitator motor was ring-mounted and designed for milk tank application. A built-in pod-type sump insured positive and complete flooding of the outlet valve down to the last cup of milk. The broad elliptical section of the tank's bottom assured maximum surface contact and built-in structural strength. Fabricated "H" section saddles supported the inner tank directly to guarantee unvarying calibration of all sizes of Sunset Bulk Milk Tanks. All models were adaptable to pipeline or unit-milking operation. The Sunset Bulk Milk Tank was quality-produced, with its operating characteristics and high-quality equipment the most modern in the industry. Seeger know-how in design and manufacturing was an important factor in maintaining a high standard of quality construction.

Seeger engineers designed the Sunset Bulk Milk Tank in a variety sizes to meet the differing requirements of all milk producers. The company carried in inventory for immediate shipment several sizes, including a 100-gallon, self-contained model, 150- and 200-gallon models that were adaptable for remote or self-contained installation, and 300- , 400- , 500- , 600- , and 700-gallon coolers for remote installation. Larger sizes such as 800- and 1000-gallon models were available on special order.

Voices *from the* **Plant Floor** *at* **Seeger**

I Never Missed a Day on the Job

"For ten or twelve years I worked in the Glass Department as group leader on the night shift. My brother Joe was the foreman of the glass room on the day shift. Later I worked on production—on all of the lines, in the paint shop, on the forklifts for the Maintenance Department. I was a member of Teamsters Union, Local 827. Four or five of my brothers worked at Seeger and four of my children. I never missed a day on the job. I got hurt on the job. While welding a 2,500-gallon milk cooler, my back went out. When I was on the night maintenance shift, I started cars on cold winter nights and plowed snow."

Louis Kuchenmeister

This photo shows how the agitator was positioned in a bulk milk cooler. It also shows the polished interior finish of the cooler. From a Sunset Equipment Company catalog in the author's collection.

Eight coolers were assembled per day. The final assembly work was done on two shifts by a work force of thirty-five employees. A small third shift undertook to provide essential machine welding and belt sanding. The slowness of several operations forced this procedure to provide material for the two assembly shifts. The U.S. Steel Corporation was one of the companies that supplied the stainless steel that was used in these tanks.

Actual production of the milk coolers began in the Metal Shop where the parts were fabricated by that department's machine operators and helpers. There, too, a tape-like adhesive material was added to the easily marred stainless steel for protection during the production process. Never before had it been necessary to handle material so carefully. Appearance played an important part in the salability of the product. This expensive tape was not removed until the necessary metal finishing operations were completed and the cooler was ready for testing and crating. A separate group of employees worked on sub-assemblies in the Drop Tank Department. The parts involved were the covers, bridges, gauge bars, and agitators.

The automatic seamwelder machine seen here was a key piece of equipment in making bulk milk coolers because it was used to build the positive flowing cold evaporator. From a Sunset Equipment Company catalog in the author's collection.

The two most important phases of the assembly work were the welding and metal finishing operations. There were three types of welding used: shielded-arc machine welding: shielded-arc hand welding: and arc welding. In the shielded-arc process, one of three possible gases (argon, helium, or an argon-helium combination) was used in this type of welding. Shielded-arc welding prevented the discoloration of the steel, reduced metal warpage, and aided in the penetration of the material. In the straight arc welding process, no gas was used. For either arc or shielded-arc, the material could simply be fused together or in cases where a cavity or gap existed, rod or wire was used. The metal finishing operations depended on a light and steady touch. Weeks were spent in trying out and acquiring the proper abrasives to provide the desired finish after the welding had been completed.

The first step in the assembly of the cooler depended upon its size. The liners and exterior shells of all coolers under 300 gallons capacity arrived in the department in one piece. When the line had to assemble a cooler that was 300 gallons or more, both the liners and the exterior shells arrived on the line in two pieces each. The halves had to be seam-splice welded together to form the shell or liner, as the case may have been. The automatic shielded arc-welding machine that completed the process was capable of handling the largest size tank Seeger produced.

From this point on all coolers regardless of size followed the same assembly procedure. The next stop for the liner was the end-welding operation. The semicircular end pieces were hand welded with shielded-arc welding equipment while being held in place on a vertical jig. A horizontal jig accomplished the same purpose for the exterior shell ends. These, however, were tack-welded in place before the welding was undertaken. With the ends in position, the parts were removed from the jigs and the corners on the liners hand welded. Coil fasteners, which helped give the tanks greater rigidity and were similar in shape to an angle iron, were welded to the side of the liner.

The next step for the liner was the beginning of a series of metal finishing operations. The interior had to be very carefully finished. A rough surface left during this or succeeding operations could have caused milk spoilage as the milk was agitated in the tank. The liner then moved to another metal finisher who completed the corner.

The bridge, a section on top of the liner designed to carry the agitator motor and the cover hinges, was the next part to be added. It was carefully positioned, held in place by jigs, and welded. The drain was also added and welded.

An unusual test was then employed on the liner. It was a red dye that had an uncanny ability to penetrate and was used to check for porosity or pinholes. If the dye appeared on the wrong side of the seam, that seam had to be rewelded and refinished until leakage was impossible.

In the meantime, the coil forcing operation took place. Various small tubing parts were added and the entire assembly was checked for leaks. This was accomplished by charging the tubing with air pressure and submerging it in water as you would an inner tube. When both the liner and the coil successfully passed their tests, assembly operations continued. A copper sheet was placed over the liner exterior for increased contact. The coil assembly was bent into shape and secured to the side brackets. Thermalmastic was then applied for added conductivity of the refrigerant to the liner.

Over time Seeger engineers introduced a second method for carrying the refrigerant. It was a laminated coil system. A solid sheet of stainless steel

A finished bulk milk cooler. The external thermometer and the drain cock are visible on the end of the cooler. The electric motor that turns the agitator is mounted on the top of the cooler. From a Sunset Equipment Company catalog in the author's collection.

was stamped with a series of convolutions or grooves to form a coil system. It was then attached by welding directly to the liner exterior. This method eventually supplanted the copper tubing arrangement. No matter which method was used, the next step was to add the fiberglass insulation. At this point, the exterior shell, which was ready for further assembly, and the liner met. The two parts were assembled and welded together. When the welding of the entire exterior had been completed, metal finishers again took over. The tape was then removed and the cooler cleaned.

The controls necessary to the proper functioning of the cooler were installed in the next step. Then a very detailed inspection was conducted. Following this inspection, the tank moved to the calibration station, a completely new and different operation. The first step in this operation was to fill the cooler to maximum capacity with water. This tended to settle the cooler for accurate measurements and removed any buckles in the liner that may have developed during assembly operations. The water was then removed so that the slow and laborious process of measuring each five gallons of capacity began. Two items were extremely important to each cooler, the dipstick and the calibration chart.

Water was added to the cooler in five-gallon increments by means of a special measuring device. After each addition the dipstick, which was nothing more than a yardstick or rule, marked in inches and subdivided to sixteenths, was used to measure the exact level of the water. Each five-gallon reading was recorded. From this information, an office clerk then prepared a calibration chart. The chart converted the measurements from inches to gallons. For coolers up to and including 300 gallons capacity, a breakdown was made to each sixteenth of an inch. The process used was one of interpolation or approximation that was done from specially prepared tables. The larger coolers were broken down in thirty-seconds of an inch. The system was accurate to within less than a gallon of the actual measurement. Because it was impractical to mark the dipstick itself in small enough gallon-sized divisions or increments, Seeger engineers had developed this conversion method as an effective, accurate and easily handled alternative.

On the measuring of the tank's capacity was completed, three copies of the conversion chart were prepared. Two were shipped with the cooler and one remained on file with the company. The cooler, the dipstick, and the conversion chart were all assigned a serial number. The three items remained together for accurate results. In this manner the purchaser could measure the contents of the cooler with the dipstick and easily use the chart to determine the volume of milk in gallons in the tank. Thus, for example, so many inches on the dipstick converted to 98 gallons of milk in the cooler. The final operation was, of course, the crating and shipping of the product.

Seeger also began making another product in 1950, commercial icemakers.[7] By 1956 the St. Paul plant had manufactured 12,187 units and shipped 11,593. A decade later the numbers increased to 23,431, and 23,993, respectively. When demand for an icemaker that could be used in private homes, bars, patios, and so forth began to grow, Seeger managers developed the Model 50 icemaker. This icemaker was essentially a reduced-sized version of the large commercial models the company was already making and selling.

Inspection

All Seeger refrigerators, freezers, ice-making machines, bulk milk coolers, and other products were inspected for quality at many steps during the manufacturing process.[8] During each day's activities throughout the plant many contacts were made with personnel of the Inspection Department. The purpose of the Inspection Department was first to maintain the high quality of the products manufactured and thus to provide customers with absolutely satisfactory consumer acceptance, and second to help reduce the overall manufacturing cost of these various products by reducing the amount of defective workmanship.

Inspectors check refrigerator shells for defects as the shells move along the conveyor line. From a Seeger News *in the author's collection.*

The Seeger policy regarding its products was: "Quality workmanship must be the guiding influence of each and every workman along the manufacturing route of our product's development—quality cannot be put into the product at the Inspection Stations." With that in mind, each of the employees had an important part to play in the building of Seeger quality products. Each of the production supervisors had an important job in the indoctrinating, training, and supervising of employees so that they would be constantly quality-minded.

The best results in quality control could be achieved when the employees understood the functional responsibilities of the Inspection Department. That department was responsible for seeing that all outside purchased materials, both raw materials and manufactured items, were acceptable for use within the specifications set up by the Engineering Department, and that they came within the acceptable quality level. This department was responsible for seeing that all manufacturing, enameling, painting, assembling, operational testing, packing, and crating operations were performed in accordance with applicable specifications established by the Engineering Department.

The Inspection Department was also responsible for the constant exercise of good judgment by all its personnel in their decisions of acceptance of outside purchased materials and manufactured items. They had to use the same good judgment in their inspection of Seeger products as manufactured within the plant to assure customers that each Seeger appliance would serve the purpose for which it was intended and that it was of good visual appearance and looked like new quality merchandise. The Inspection Department did not actually establish all of the standards of quality for which they inspected the products, but it had the responsibility of inspecting for many specifications set up by the Engineering Department. Many of these specifications resulted from specific requirements that were inherent in the product itself because of its nature and the purpose that it had to serve. Frequently these details came directly from customers or were the direct requests of the Sales Department. Other specifications, however, were necessitated by the particular type of manufacturing equipment that had to be used in the manufacturing process, while still others were the result of the use of certain materials in the manufacturing process at the time. It is not difficult to realize that each and every employee had a part to play in maintaining the quality of the products.

The functions of the Inspection Department began with the pilot model of each newly designed refrigerator that came out of the Engineering and the Research and Development Departments. The pilot model was thoroughly inspected for the purpose of determining that it would be built according to the Engineering Department's drawings and specifications and that it was up to the quality level that would be acceptable to customers. Each die-made part was carefully and thoroughly checked before the dies were approved for production. Each outside purchased item was carefully and thoroughly inspected to see that it met the quality level that would be acceptable on the product. There were specifications pertaining to plating, hardness, durability under humidity conditions as well as many other atmospheric conditions, taste and odor, acid resistance, compression, stain, bonding, aging, tensile properties, moisture resistance, strength, and many others that were checked by laboratory technicians.

In the plant the Research and Control Laboratory performed these functions. The Inspection Department was responsible for submitting to the Laboratory all such materials and parts and finished products as required for these tests. It was also the responsibility of the Inspection Department to work with the vendors, through the Purchasing Department, to get and keep all such specifications in control. Any such specifications that Seeger's own

After the initial metal finishing work has been done on these refrigerator shells, inspectors check them for any flaws in the metal. Small defects were repaired. Shells that had serious flaws were set aside. From a Seeger News *in the author's collection.*

manufacturing processes were required to meet were similarly checked on and kept in control. Continuing with the pilot model, all fabricating, metal finishing, assembling, porcelain enameling, painting, packing and crating operations were checked to make sure they were according to drawings and specifications. The completed product was subjected to operational running tests, if the unit was to be installed in the Seeger factory. It was also subjected to various humidity, taste and odor, and seal tests. In addition it was exposed to shaker and drop tests, simulating transportation conditions, for the purpose of seeing whether or not there existed any constructional weaknesses or packing and crating failures.

After the Inspection, the Sales, the Production, the Engineering, and the Research and Development Departments had approved the pilot model, the Purchasing Department proceeded to secure the raw materials and outside purchased items. The factory would then start the manufacturing of the product. Once manufacturing operations for a specific product had reached this stage, the functions of the Inspection Department become those of quality control. All incoming purchased items were cleared through the Receiving Inspection Division of the Inspection Department. Here they were checked by lot sample for purpose of determining whether or not the part was up to acceptable quality level and that random samples were submitted to the Laboratory for the various tests covering material specifications. If the lot sample check indicated rejections beyond the allowable quality level, then the shipment would be subjected to a 100 percent detail inspection, or returned to the vendor. This decision was reached by coordination with the Purchasing Department.

All of the manufacturing processes and operations within the factory were kept within the quality level by the inspections conducted at the various Inspection Stations through which the individual parts and completed assemblies passed. Sheet steel was inspected as it was being loaded from the steel storage warehouse for the shears or presses. By this control, steel that might have caused fabricating difficulties or rejections in the finishing stages was prevented from getting into production channels. Prior to production of parts on the various presses, the first parts off the dies were checked by production supervisors and then by collaboration with the Inspection Department to determine whether or not the part as produced met its specifications. If the part was acceptable, production continued to run. If the part was rejected, then the Tool and Die Department technicians were called in for the purpose of making necessary corrections to produce the part so it would be acceptable to Inspection. During the time parts and assemblies were being produced, the various machine operators, departmental supervisors, and floor inspectors made regularly scheduled checks to see that control was being maintained.

While the parts and assemblies progressed through the Metal Fabricating and Metal Assembly departments, they were inspected 100 percent for general conditions and all metal conditions prior to passing on into the Porcelain Enameling or Dulux Paint Departments. In these departments all parts and assemblies were inspected 100% for conditions of the porcelain or paint, as the case may have been, prior to their acceptance for the assembly lines. It was the responsibility of the Inspection Department to send random samples of porcelained and painted parts to the Laboratory at regular intervals for tests on acid resistance, humidity, film thickness, and so forth. These tests were all conducted for the purpose of keeping all processes within control. It was also a function of the Inspection Department to make a check on the film thickness on all painted parts during each shift of production. Proper film thickness was a very important factor in the durability, corrosion resistance, and visual appearance of the paint. Thus quality control of the film thickness was critical.

As the refrigerators and freezers were assembled on the various assembly lines each operation performed was checked on by Inspectors located at various stations along the course of the assembly lines. The insulating of the cabinets was carefully checked as was the vapor sealing. These requirements were highly important to the efficiency and satisfactory performance of the refrigerator. All materials and parts used in the assembly were inspected to determine whether or not they had been damaged in any way through transporting to the lines or while being assembled. Also, the Inspection Department made careful checks to make sure that the materials and parts used in the assembly were those intended for the model being produced. Unit running tests and leak tests were made on each refrigerator in accordance with established specifications, as also were light, temperature control, thermometer, and seal tests. Again, each and every one of these tests was conducted for the purpose of insuring that the unit and all its mechanical and electrical parts functioned in such a manner as to give the customer the utmost in refrigeration efficiency.

One of the most important tests for refrigerators was the leak test, which was conducted to determine if there was a refrigerant leak. All shelving and accessories were inspected for proper fit, operation, appearance, and packing. All hardware was inspected for operation, fit, and appearance. The door seal, alignment of panels, and general appearance of the interior and exterior of the cabinet was also carefully checked. All packing and crating was inspected so as to help keep these operations in control so they met specifications.

At regular intervals the Inspection Department submitted completed refrigerators to the Laboratory for their operational and odor tests. They also conducted shaker and drop tests on completed refrigerators at regular intervals. In addition they would regularly set up and super-critically inspect completed crated refrigerators that had been selected at random from production lines. The purpose of these tests and inspections was to help keep all manufacturing processes in control. Another function of the Inspection Department was to control the proper segregation, reporting, and disposition of any outside purchased materials or parts that for any reason were unusable in the manufacturing process. This was accomplished through close coordination with the Production Departments and the Purchasing Department.

A final responsibility of the Inspection Department was to investigate any report received that indicated that the product had not met with satisfactory customer acceptance after it had reached its destination. Depending upon the results of such an investigation, steps were taken by one or more

Seeger Refrigerator Company Facilities, 1950

LOCATION	GROUND SPACE (ACRES)	FLOOR SPACE(SQUARE FEET)
St. Paul (Main Plant)	10.2	551,976
St. Paul-Gloster*	28.3	73,271
Evansville (Main Plant)	40.0	878,128
Evansville (Plant No. 3)	2.0	51,500
Evansville (Plant No. 6)	1.5	73,972
Totals	82.0	1,628,847

*The Gloster facility was just outside the city of St. Paul in what is today the city of Maplewood. It was primarily used for warehousing and was bounded by Frost Street, English Street, Good Avenue, and Frisbie Avenue.

This page from a June 1945 Seeger News *carried aerial photos of the Arcade Street plant in St. Paul (top) and the main plant in Evansville (bottom). From a* Seeger News *in the author's collection.*

Population, Wages, and Income between 1900 and 1960

Over more than five decades the Seeger Company introduced and embraced the latest advancements in refrigeration engineering and manufacturing operations. Riding the crest of demand for commercial and residential refrigeration products, the company's strategic decisions reflected a keen commitment to entrepreneurship, vision, and management ability. An expanding U.S. population, which continuously added potential customers to the marketplace for Seeger products, increased from 75,994,000 in 1900 to 179,323,000 persons in 1960. Population growth, especially after 1950, helped fuel sales growth at Seeger. This population increase is summarized in the following table.

U. S. Population, 1900–1960

Year	*Number of Persons*
1900	75,994,000
1910	91,972,000
1920	105,710,000
1930	122,775,000
1940	131,409,000
1950	150,697,000
1960	179,323,000
2006	300,000,000

In addition to an expanding population, the Roosevelt Administration's Rural Electrification Act (REA) of 1935 used federal lending authority to help bring electricity to isolated rural areas, most of which were not served by private utilities. Wider availability of electrical power significantly increased the demand for refrigeration products in the countryside and small towns of America. The 1940 U.S. Census was the only national survey to determine the impact of the REA by tabulating the number of dwellings with an electric refrigerator. Statistical analysis of the impact this federal program had on rural areas of the nation is hampered because no other federal survey data is available for comparison purposes.

The destruction of personnel records of the Seeger Company in 1984 limits our knowledge of the wages, salaries, and benefits paid to its employees throughout the company's long history. Anecdotal evidence supplied by retired and former employees confirms that certain benefits were available and a Seeger Employee Handbook, from about 1946–50, details the specific programs the company offered at that time. Regretfully, it is not possible to reconstruct the year that specific benefits such as life insurance or a retirement pension plan were introduced at Seeger. U.S. Commerce Department data does provide, however, the average hourly earnings for manufacturing employees from one decade to the next. These federal reports indicate that the average hourly earnings for manufacturing employees gradually increased between 1909 and 1970 from about 19¢ per hour to $3.81 per hour. This information is summarized in the following table.

Average Hourly Earnings in Manufacturing

Year	*Hourly Earnings*
1909	$0.19
1919	$0.48
1929	$0.57
1937	$0.62
1945	$1.02
1957	$2.07
1970	$3.81

In general rising hourly wages in a nation encourages greater levels of consumption. Thus over the first fifty or so years of the Seeger Company's business life, more people in the United States occupied more homes and farms that had electricity and over time were earning more per hour which provided additional income to purchase household appliances such as refrigerators, freezers, and ice machines that were made by Seeger employees.

of the following departments to make such changes as were necessary to correct the condition on future production: Engineering, Inspection, Production, Purchasing, Research and Development. Thus, the high quality level of Seeger products was maintained.

By 1950 the Seeger Company, with plants operating at full capacity in St. Paul, and Evansville, Indiana, manufactured a line of domestic and commercial refrigeration units, freezers, and dehumidifiers at the rate of 815,000 units per year. The company employed more than 6,100 people, had over 1,600,000 square feet of factory floor space and total assets in excess of $35,000,000 (the equivalent of $271,313,453 in 2005 dollars). Sales of refrigeration products to Sears Roebuck and Company, Montgomery Ward and Company, Admiral Corporation, and Frigidare were over $80,000,000 in the fiscal year 1950 (the equivalent of $620,145,037 in 2005 dollars). The table on page 164 sums up how sizable the Seeger Company's operations had become by 1950 as a consequence of its central role in the manufacturing of a wide range of residential and commercial appliances for a national market in the United States.

At mid-twentieth century the company also witnessed a signal transition: it experienced the passing of executive leadership from the second generation to the third generation of the Seeger family when John S. Holl succeeded Walter Seeger as president. Walter Seeger, continued to exercise strong and active involvement in corporate affairs as chairman. At this time Neil H. Griebenow was the first vice-president and general manager; Russell Ayres, the chief engineer; John W. Kreuger and John W. Seeger were vice presidents; Clarence W. Moberg was the secretary and treasurer; Walter A. Holt was assistant secretary and assistant treasurer; L. Kern Sosey was works manager; Walter Ouderkirk was plant superintendent; Irving Warren was assistant secretary; Frank B. Taylor was the personnel and public relations officer; Philip Bourne was purchasing agent; and Paul Bowman was mechanical superintendent of maintenance.

Over the years from 1946 to the merger with Whirlpool and RCA in September 1955, Seeger generated $710,666,217 in gross sales and $37,094,374 in net income, the equivalent of $4,882,571,910 and $1,606,586,745, respectively in 2005 dollars.[9] According to the financial publication *Investor's Reader* (April 16, 1947), only in 1930 had the Seeger Company incurred a loss for its fiscal year. Thus the initial postwar decade had proved to be a period of extraordinary productivity and profitability for this company located on St. Paul's East Side.

10

The 1955 Merger with the Whirlpool Corporation and the RCA Appliance Division

THE DECISION TO MERGE the Seeger Company with the Whirlpool Corporation, the laundry equipment manufacturer of St. Joseph, Michigan, and the electric stove division of the Radio Corporation of America grew out of the way the home appliance industry had evolved during the first half of the twentieth century. In 1900 in nearly every town or city of any size there were skilled artisans, cabinetmakers, and metal workers who crafted for the local community iceboxes, washing machines, and stoves. During the passing of the decades many of these independent entrepreneurs retired from business, died, or combined their production facilities with similar local firms. The consolidation continued at a gradual pace in the industry locally, regionally, and nationally during the 1920s, '30s, and '40s. By the 1950s the tempo of such mergers had increased in many industries including automobiles, steel, railroads, and natural resources. The home appliance manufacturers were ripe for such action because a merged manufacturing and marketing operation would provide greater profitability from production economies, reduced personnel costs, and an increase in the number of products that could be manufactured and marketed under a common brand name.

Walter Seeger recounted to the author that it was General Robert E. Wood, the chairman of Sears, Roebuck and Company, who first suggested at least by December 1954 that two of his company's leading appliance suppliers, Seeger and Whirlpool, should consider merging their firms. Wood's suggestion must have received serious consideration from Seeger as the men had been close business and personal friends for more than twenty-five years. Together they had planned and launched the first *Coldspot* refrigerator in 1930 and the first Sears home freezer in 1939. Elisha Gray II, the president of the Whirlpool Corporation in 1955, recalled in a conversation with the

Walter G. Seeger on the Merger

I am enthusiastic about the prospect of the new company that has been formed by combining Seeger, RCA-Estate, and Whirlpool.* The possibilities for added sales volume in our present product lines and adding new products are very good. The products we have now are of a quality and design that fully justify the enthusiasm of our customers. The combined resources of our new company should enable us to make these and the new products even more attractive to housewives than they are today.

The outlook for the future, in turn, makes the prospects for more jobs, steadier work, and greater opportunities for advancement in the whole organization even better than they have been.

No matter how good any company organization looks on paper, however, its success depends on people. And this is one of the biggest reasons for my enthusiasm about the new company.

During the discussion and planning session that led to the merger, I visited the other plants. One of the things I found during these visits was the way the people in the different parts of the organization seem to be together. Often during these visits I would think to myself, "these are our kind of folks." Very often, while I was thinking that, someone from another plant would say something suggesting the same idea about Seeger people.

Sometimes a merger is likened to a marriage. The prospects of a long and happy life for this merger seem good to me because I feel the people are especially compatible. There is one more similarity I would like to point out: No matter how good its prospects, a merged company, like any other, will work only if we all do our best to make it work.

*1956 *Annual Report Whirlpool-Seeger Corporation* (St. Joseph, 1956): 4.

author (in 1995) that the merger was undertaken because the participants "liked each other." While that elementary comment may be interpreted from several perspectives, the available evidence suggests that it was simply a sustaining propellant for the discussions and does not seem to have been the driving engine.

There were compelling economic reasons within the appliance industry that favored the proposed merger. Trends in the industry clearly indicated that success depended upon extensive research programs, the application of the most advanced technology, low-cost manufacturing processes, and comprehensive and aggressive distribution of products. These requirements could best be financed and realized through a larger volume and broader product lines that only the merged company could offer. Furthermore, Walter Seeger, nearing seventy years of age, needed to consider the interests of his shareholders. The family's fifty-three-year-old firm could not continue to maintain its historical presence as manufacturer of only refrigerators and freezers in a national marketplace that was rapidly changing because mass merchandisers such as Sears and Montgomery Ward had gained increasing influence over consumer buying choices due to their marketing, advertising, and retailing clout. Realistically, changing times prompted Walter Seeger to examine how his company could broaden its line of products. Since 1930 the firm had been tied primarily to one major customer—Sears, Roebuck and Company. In fact, in 1955 more than ninety percent of the company's production was earmarked for the mass-market merchandiser.

Although the situation was similar at Whirlpool, that company's financial position was complicated by heavy debt, partly generated by the expensive acquisition in 1952 of a basic supplier, the Clyde Porcelain Steel Company of Clyde, Ohio, and a factory in Marion, Ohio, in early 1955. In the wake of these acquisitions, Whirlpool began actively searching for a substantial and immediate infusion of cash. A confidential internal memo circulated among Whirlpool's participants in the merger discussions disclosed the company's urgent financial situation. In order to survive, the struggling Whirlpool Corporation needed cash; and the opportunity to join in a merger with Seeger and RCA's

Delaware Appliance Division would bring substantial funds to a new corporation. As part of their due diligence work as the merger talks progressed, Whirlpool's representatives learned that Seeger had $28 million in reserves and RCA was prepared to contribute $16 million, a total of $44 million, to the creation of the new merged corporation. A capital infusion of this scale would definitely bring aid to Whirlpool's dire circumstances.

Walter G. Seeger. Photo courtesy of John S. Reay.

Although possible combination discussions had been underway since December 1954, the first mention of the plans occurs in the Minutes of the Meetings of the Seeger board of directors of May 3, 1955.[1] The minutes state that Seeger board member Robert E. Brooker, a vice president of Sears, Roebuck and Company, reviewed the status of the negotiations he had had with representatives of the Whirlpool Corporation and the Radio Corporation of America regarding a possible merger. The details of the proposed terms for the merger that Brooker summarized for the board were in fact the ones that the three companies finally agreed to when the merger was completed.

Shareholders of the company were advised in a letter dated July 16, 1955, of the plan to form a new company that would acquire, own, and operate the businesses carried on by Seeger, Whirlpool, and the electric stove and air conditioning subsidiary of RCA.[2] The plan contemplated that Seeger and Whirlpool would be merged with RCA's Delaware Appliance Corporation to which RCA would transfer certain assets in the net amount of $21,600,000, consisting of cash and the assets of its stove and air conditioning business.[3] In addition, RCA would enter into an agreement covering the use by the merged company on its products of the *RCA* trademark in combination with the *Whirlpool* trademark.[4] The corporation, as merged, would be known as the Whirlpool-Seeger Corporation.[5]

Elisha Gray II, the President of the Whirlpool Corporation

President of the Whirlpool Corporation at the time of the 1955 merger with Seeger and the RCA Appliance Division, Gray was born in New York City on September 7, 1906. A graduate of the Massachusetts Institute of Technology, he began his career at the Sears home office in Chicago in 1928. Rising through a number of merchandising positions, he left Sears in 1938 to join the Nineteen Hundred Corporation in St. Joseph, Michigan, to maintain liaison with Sears which was then purchasing most of the company's production of home laundry equipment. He rose through the ranks of executive positions becoming president of the Nineteen Hundred Corporation in 1949. In 1950 Gray's company changed its name to the Whirlpool Corporation. He became president of Whirlpool-Seeger in 1955 and chairman of the Whirlpool Corporation in 1958.

Seeger

The Seeger Company was a leading national manufacturer of electrical refrigeration equipment. Since 1930 the firm had manufactured home electric refrigerators and since 1941 of freezers for sale by Sears, Roebuck and Company under its trademark *Coldspot.* From the late 1940s, Seeger also manufactured dehumidifiers for sale by Sears under its trademark *Coldspot,* as well as manufacturing a line of commercial refrigerators and ice cube makers for the Frigidaire Division of the General Motors Corporation.

During World War II, the Korean War, and until 1955, Seeger had produced aircraft equipment as part of the defense program. Defense sales, which amounted to about 38 percent of total sales for the fiscal year ended August 31, 1954, had decreased after that time because of the termination of some contracts and the completion of others so that in July 1955 defense sales amounted to only about 16 percent of total sales.

In addition, the Seeger Company had for many years manufactured rotary-type compressors for refrigeration equipment. It had designed in the early 1950s a line of larger rotary-type compressors suitable for freezers and room air conditioning equipment. Production of these larger compressors had been limited to experimental units for field-testing purposes. These larger sized units enabled Seeger to offer a line of compressors to other manufacturers of freezers and room air conditioners.

In 1954 Seeger had also established a program for engineering and tooling to manufacture refrigerated bulk milk storage tanks for farm use. These storage tanks, the exterior and interior of which were made of stainless steel, opened up a new field for refrigerated products. Seeger subsequently manufactured bulk milk storage tanks at the St. Paul plant for more than twenty-five years. This equipment was sold under a private brand name for the Sunset Corporation of St. Paul.

At the time of the proposed merger, the Seeger firm manufactured the principal components of its products and assembled the finished products in combination with parts that it purchased from other sources. Consequently the company produced the refrigeration units for home refrigerators and for freezers up to 14-cubic-foot capacity, while refrigeration units for refrigerators and freezers with more than a 14-cubic-foot capacity were purchased from other firms.

Sales of the *Coldspot* line of home refrigerators and freezers to Sears accounted for most of Seeger's total sales. For the fiscal year ended August 31, 1954, sales of

Elisha Gray II on the Whirlpool-Seeger Plants

We have several plants that, I dare say, are as efficient as any in the country on somewhat of a job lot basis.* Probably the greatest assortment of products we have comes out of our St. Paul plant. It's by no means our largest facility, but it's just as efficient as it can be on the manufacturing of relatively small numbers of commercial products.

**Appliance Manufacturer* (Chicago, [January] 1957): 43.

these products to Sears represented approximately 90 percent of all non-defense sales.[6]

For many years Seeger had developed and maintained a first-class research and engineering staff and had retained designers whose activities were directed toward improvements in products and production methods. The company's general offices and one of its plants were located in St. Paul, Minnesota, and a second production facility was located at Evansville, Indiana. In August 1955 the buildings of both plants were in good repair and had been improved in recent years. Between 1950 and 1955 the Seeger management had spent approximately $4,300,000 on the two plants and on the purchase of new or improved equipment. The Evansville plant primarily manufactured refrigeration units and home refrigerator cabinets and assembled home refrigerators. In St. Paul, the factory primarily constructed freezer cabinets and other components and installed refrigeration units in freezer cabinets. Defense work was carried on at both the St. Paul and Evansville plants. In addition to owning these properties, Seeger also leased space near the St. Paul and Evansville plants totaling 340,000 square feet. In August 1955 Seeger employed approximately 5,000 persons.

Whirlpool

The origins of the Whirlpool Corporation are complex. In 1929 the Nineteen Hundred Corporation, a predecessor company of the Whirlpool Corporation, had been incorporated under the laws of the State of New York. This was the surviving corporation of the merger between the Nineteen Hundred Washer Company, founded in 1898, and the Upton Washing Machine Company, which was founded in 1911. This merged firm was known as the Nineteen Hundred Corporation until 1950 when the name was changed to the Whirlpool Corporation in a marketing strategy orchestrated to promote the company's independent brand of laundry equipment.

The word "Whirlpool" in the new corporate name of the former Nineteen Hundred Corporation was of recent vintage. Beginning in 1950, the company had used the name "Whirlpool" only in reference to the predecessor laundry appliance manufacturing company. Historically, the term was first used about 1930 to identify as a brand name a washing machine manufactured by the Nineteen Hundred Corporation of Binghamton, New York, and St. Joseph, Michigan. The Nineteen Hundred Corporation was, as mentioned, the result of the 1929 merger with the Upton Washing Machine Company of St. Joseph, Michigan. By 1950 the management of the Nineteen Hundred Corporation recognized that it needed to closely identify itself with the brand name of a product it manufactured. So it

In 1957 passersby in downtown Minneapolis could see the RCA Whirlpool Miracle Kitchen that was displayed in one of the windows at the Dayton's Department Store on Nicollet Avenue between Seventh and Eighth streets. Norton & Peel photograph. Courtesy of the Minnesota Historical Society.

changed its name to the Whirlpool Corporation. To represent to the public that today's Whirlpool Corporation had its origins exclusively in the St. Joseph, Michigan, laundry-appliance manufacturing company is a distortion of the present-day Whirlpool Company's rich and vital history (see Appendix C for the Genealogy of Whirlpool Business Interests). Such a limited view overlooks in an exclusive and parochial manner the complexity of Whirlpool's development through mergers and acquisitions.

In 1955 Whirlpool was the leading manufacturer in the United States of home laundry equipment in a field crowded with such widely recognized competitors as Maytag, Bendix, Westinghouse, General Electric, Norge, Speed Queen, and Frigidare. Unlike the refrigeration industry that had experienced extraordinary technological and design advancement between 1900 and 1940, laundry appliances had lagged in both respects until after World War II. With the introduction of improved products after the war, sales of laundry appliances increased dramatically in the late 1940s and 1950s.

Sales of the *Whirlpool* line of appliances had shown a steady increase since their introduction. In 1954 they represented approximately one-third of the company's total sales of laundry equipment. Unlike the products of Seeger, the *Whirlpool* line was sold to appliance distributors who, in turn, sold to about 15,000 retail dealers throughout the United States. At the time of the merger, sales of the *Whirlpool* line were made in limited volume in the export market. Whirlpool had also entered into contracts with manufacturers in Canada and several other foreign countries under terms to furnish engineering and manufacturing assistance in the design and manufacture of laundry equipment for fees based on sales.[7]

Whirlpool's products were produced in plants in St. Joseph, Michigan, and Clyde, Ohio, both from parts made by it and from those purchased from others. At the time of the merger with Seeger and RCA, Whirlpool's management anticipated that its relationship with Sears would in the future, as in the past, be profitable.

During the months of discussion about a merger, Whirlpool's products consisted of automatic washing machines, automatic gas and electric dryers, wringer washing machines, and rotary ironers (commonly known as mangles). For many years Whirlpool had been the sole supplier to Sears of its *Kenmore* line of home laundry equipment. Prior to World War II, substantially all of Whirlpool's laundry equipment business consisted of sales of the *Kenmore* line to Sears. After World War II, Whirlpool began the manufacture and distribution of a separate line of laundry equipment under the *Whirlpool* brand name.

While dollar sales of the home laundry equipment industry had grown about 69 percent after 1947, the first full year following World War II, Whirlpool's total sales of such products had increased from $23,867,000 in 1947 to $156,065,000 in 1954, an increase of 554 percent. This growth had been due in large measure to the replacement of pre-World War II wringer-type washers by automatic washers, the growing consumer acceptance of automatic dryers, and such strong economic factors as increases in population, families, home construction, and spendable income. In August 1955, Whirlpool had approximately 5,700 employees.[8]

RCA's Delaware Appliance Corporation

Since 1952 RCA had manufactured and sold electric and gas kitchen ranges under the brand name *RCA Estate* through a subsidiary, the Delaware Appliance Corporation. The company also sold room air conditioners manufactured by the Fedders-Quigan Corporation (this was under a contract that was to terminate on June 30, 1956) under the name *RCA*. RCA was not a

These two pages are taken from a 1957 Whirlpool-Seeger catalog of their home appliances. The top illustration shows the Imperial Refrigerator-Freezer, which featured the RCA Whirlpool logo near the top of the refrigerator door. All the features and specifications for this appliance were supplied on a separate page (bottom). The 1957 catalog was printed in color so that potential buyers could see that they had a choice of "Decorator Colors" (white, pink, yellow, or green) for the appliance's exterior. One message that this catalog conveyed to consumers was that economic hard times and food rationing were no longer an issue in American homes. From a Whirlpool-Seeger catalog in the author's collection.

RCA WHIRLPOOL Imperial Combination Refrigerator-Freezer

model DI15B . . . 15.5 cubic foot . . . 166 pound freezer

two appliances in one

A big automatic defrosting refrigerator section and a big true zero-degree freezer . . . each with their own door and both separately insulated and refrigerated. "Right-Side-Up" . . . the refrigerator section is on top where it's most convenient.

New, Exclusive Air Purifying System keeps food fresher, longer. Lets foods — even whipping cream and onions — live happily together, ends "refrigerator smell" forever. The Purifying Lamp ejects purifying ultra-violet rays that attack air borne bacteria, retards mold growths and prevents the transfer of odor and taste which contaminates food. *The balanced cold Air Purifying System* distributes the purifying air evenly throughout the refrigerator section . . . all the air that touches your food is sanitized every 45 seconds . . . and an ideal food preserving temperature is maintained in all parts of the refrigerator — even the door.

Completely Automatic Defrosting refrigerator section ends defrosting worries forever. No buttons to push . . . no pans to empty.

Ice Water Tap (optional at slight additional cost) . . . just press the glass against the control for ice water. Refills automatically as water is withdrawn.

3 Glide-Out Shelves bring food right out in front. Convenient . . . strong . . . easily removed completely for cleaning.

Deep, Roomy Door Shelves hold half-gallon milk containers, tall bottles, often used items . . . all at perfect temperatures.

Butter Conditioner has individual temperature control. Keeps up to one pound just right for easy spreading.

Two Egg Compartments with removable egg racks that can be carried where needed. Hold 18 eggs . . . 9 in each rack.

Cheese Keeper stores up to 2 pounds right at your finger tips. Removable cheese dish included.

Bacon Keeper holds 3 pounds at safe temperatures. So handy!

Big Window Fruit Bin lets you see what's inside. Tilts out for complete accessibility. Holds up to 22 pounds.

Twin Porcelain Enameled Crispers keep vegetables fresh and crisp. Tremendous capacity . . . holds almost a bushel.

Meat Keeper holds over 11 pounds. Located up front . . . glides out for extra convenience. Transparent so meats are easy to see.

Big Automatic Interior Light has twin bulbs . . . floods every corner with light.

Adjustable Temperature Control lets you set the temperature just where you want it.

166 pound true zero-degree freezer

Quick Freeze . . . Store Frozen Foods Up To a Year with this jumbo-size freezer section. Two prime freezing shelves for more fast freezing surface, even temperatures . . . safer food protection.

New Instant Ice Cube Ejector automatically ejects ice cubes into the handy server bin by just pressing lightly on the handle. Server bin holds up to 200 cubes . . . 8½ pounds. 4 ejector trays . . . each holds 20 of the new slim cubes that fit into small glasses, cool faster.

Automatic Ice Cube Tray Filler (optional at slight additional cost) refills tray right inside the freezer section. Just place the tray under the filler and touch the control . . . the tray automatically fills perfectly.

Full Width Freezer Basket glides out smoothly and easily on nylon slides . . . even when fully loaded. Holds 40 pounds.

Two Freezer Door Shelves are perfect for frozen juices, soups, small items. Shelves tilt out for accessibility.

beauty, quality, dependability

Decorator Colors . . . Choose from pink, yellow, green or white exteriors. Interior is Dawn Pink, accented with Dusk Pink, Midnight Charcoal, gold and chrome.

Vacuum-Tested Cabinet assures there are no air leaks to affect food temperatures . . . that moisture laden air won't seep in to ruin insulation.

Advanced Design Refrigerating System is hermetically sealed. New extra-capacity rotary compressor is built for economical, long-life, heavy-duty service. Sealed in steel, it never needs oiling. Spring mounted for quiet, vibration-free operation. Refrigerant is safe, odorless Freon. Proven dependability . . . over 5,000,000 in use.

Five-Year Warranty includes one-year warranty on complete refrigerator against defects in material and workmanship — plus an additional 4-year protection plan on the hermetically-sealed refrigerator system.

Available with Left Hand Door

SPECIFICATIONS
DI15B

Total Net Storage Volume (N.E.M.A.)	15.5 cu. ft.
Refrigerator Storage Volume	10.7 cu. ft.
Freezer Storage Volume	4.8 cu. ft.
Freezer Food Storage Capacity	166 lbs.
Net Shelf Area	27.6 sq. ft.
Height (A)	65½"
Width (B)	34"
Depth Overall* (C)	32¾"
Width with Door Open (D)	39¾"
Depth to Extreme Swing of Door (E)	59¼"
Compressor	1/5 hp.
Net Weight (Approx.)	386 lbs.
Shipping Weight (Approx.)	446 lbs.
Electrical Requirements	115-V., 60-C., A.C.

Accessory ice water tap and ice cube tray fillers are easily installed and connected to cold water supply with ¼" copper tubing. Specifications subject to change without notice.

*Doors may be removed so refrigerator will fit through 27" door.

Exclusive Air Purifying System constantly moves cold air over the Purifying Lamp, and then carries it to every corner. Foods won't trade odors or tastes . . . stay fresher, longer.

New Ice Cube Ejector removes cubes from trays easily and fast . . . cubes fall automatically into the server bin. Always a supply of ice cubes at your finger tips. Cubes will not freeze together in the bin.

Glide Out Shelves put food right out in front where it's easy to see and reach . . . over 70% of the shelf glides way out on lifetime nylon glides . . . putting even objects in the back and corners at your finger tips. Made of rustproof aluminum.

Big Convenience Door with specialized storage . . . butter conditioner, cheese keeper, bacon keeper, and 2 egg compartments . . . perfect places for those often-used foods. Door shelves are deep and roomy — hold even half-gallon milk containers . . . all at the right temperature.

RCA WHIRLPOOL HOME APPLIANCES

Form No. 3571 Products of **WHIRLPOOL-SEEGER CORPORATION** St. Joseph, Michigan Litho in U.S.A.

major domestic manufacturer of these products and was not a supplier of stoves and room air conditioners to Sears.[9] General David Sarnoff, the chairman of RCA, wanted his corporation to abandon the production and sales of these products. RCA's participation in the merger of the Seeger and Whirlpool companies to form the Whirlpool-Seeger Corporation would accomplish this goal. Furthermore, RCA was prepared to contribute a substantial sum of cash to the new corporation and allow it to use for a period of ten years with the payment of a fee the coveted *RCA* trademark on its appliances.[10] In August 1955 the stove manufacturing plant was located in Hamilton, Ohio, and had approximately 500,000 square feet of floor space. At that time the Stove and Air Conditioning Departments had approximately 700 employees.

In preparation for the merger, the Delaware Appliance Corporation acquired from RCA assets in the net amount of $21,600,000 consisting of cash (estimated at $16,000,000) and the assets (other than accounts receivable) of its stove and air conditioning businesses. The transfer included certain patent rights that RCA regarded as allocable to its air conditioning and stove operations and excluded other patent rights of possible application to air conditioners that RCA did not deem allocable to the operations. In addition, prior to the merger RCA entered into an agreement with the Delaware Appliance Corporation permitting Whirlpool-Seeger, in respect of certain specified products to be manufactured or sold by the new corporation (including washing machines, dryers, ironers, refrigerators, freezers, dehumidifiers, air conditioners, and electric and gas ranges), the exclusive right to use, as a trademark, the *RCA* trademark in combination with the *Whirlpool* trademark, the combined trademark being referred to as the *RCA Whirlpool* trademark.[11]

Following the merger RCA entered into a five-year agreement with Whirlpool-Seeger that provided for RCA to furnish Whirlpool-Seeger with advice and assistance in the organization, expansion, and operation of research and development facilities, advice and information concerning research of RCA useful to the manufacture and development of Whirlpool-Seeger products, advice and information respecting market research and economic planning and forecasting, and advice and assistance in establishing service activities with respect to Whirlpool-Seeger products.[12] For these services Whirlpool-Seeger paid RCA $86,667 per month, or $5,200,020 during the term of the agreement.

What the Whirlpool-Seeger Merger Meant for Sales

Walter G. Seeger said this about the merger with Whirlpool: "With the merger, we have a distributor setup second to none. We fully expect to stay ahead of the whole appliance business—we have to." In the first year after the merger, the new corporation set its initial sales goal at $280 million dollars. Actual sales soared to more than $380 million dollars. One of the prime objectives of the merger was a complete line of major appliances. By the end of 1956 that objective had been achieved with Whirlpool-Seeger selling a full line of home laundry equipment, refrigerators, freezers, ranges, dehumidifiers, air conditioners, and vacuum cleaners. By 1957 the company had 125 different models of home appliances.

From the 1955 Annual Report Whirlpool-Seeger Corporation

On December 31, 1955, Whirlpool-Seeger Corporation had been in operation for exactly 107 days. The Company was formed by the merger on September 15, 1955, of Whirlpool Corporation, Seeger Refrigerator Company and the Delaware Appliance Corporation. The latter had been incorporated in August 1955 to which Radio Corporation of America had, prior to merger, transferred its kitchen range and air conditioning departments.

Combining the results of operations of Whirlpool and Seeger for the period from January 1 to September 15, the date of the merger, and of Whirlpool-Seeger from September 15 to December 31, total net sales for 1955 were $280,362,765, and combined net earnings after provision for taxes on income were $14,129,908.

Adding the operations of RCA stove and air conditioning departments for the period from January 1 until September 15, total sales were $301,695,814, and net earnings after provision for taxes on income were $14,628,752. After provision for dividends on the 4⅛% Cumulative Convertible Preferred Stock for the full year, the net earnings of $14,129,908 amounted to $2.70 per share on the average of 4,970,429 of Common Shares outstanding during the year, and the net earnings of $14,638,752 amounted to $2.40 per share on the 5,804,857 Common Shares outstanding at the year-end.

Walter G. Seeger, Chairman of the Board of Directors
Elisha Gray II, President

After reviewing the merger proposal, the Seeger board of directors voted to recommend it to the shareholders for approval.[13] The combination appeared to be well suited to accomplish the distinctive purposes of the parties. It would permit the new Whirlpool-Seeger Corporation to offer a broader line of household appliances to distributors and dealers and to do it under one brand that encompassed both the RCA and the Whirlpool names that had already received a high degree of public acceptance. It would result, too, in expanded research and development activities in the home appliance field and permit more efficient use of the present plants and equipment of the constituent companies.

Sears, which was at the time a stockholder in both Whirlpool and Seeger, planned to continue as such in the merged Whirlpool-Seeger and would also continue to use the new corporation as its principal source for home laundry equipment, refrigerators, and freezers, although Sears in 1955 had no plans for purchasing stoves or air conditioners from Whirlpool-Seeger.[14]

Capitalization and Conversion of Securities

Following shareholder approval, the Seeger board of directors advised stockholders that the proposed merger would became effective on September 15, 1955, and that the Whirlpool common stock outstanding immediately prior to the merger would be converted into Whirlpool-Seeger common stock on a share-for-share basis and that the Seeger common stock outstanding immediately prior to the merger would be converted into common stock and a 4¼ percent cumulative convertible preferred stock of Whirlpool-

The Board of Directors of the Whirlpool-Seeger Corporation in late 1955. Clockwise from the head of the table, they were: Walter G. Seeger, chairman of the board (former chairman of the board, Seeger Refrigerator Company); Edward Gudeman, vice president (Sears, Robuck & Company); John S. Holl, vice president, Seeger division (former president, Seeger Refrigerator Company); Elisha Gray II, president and chief executive officer (former president and chief executive officer, Whirlpool Corporation); Frederick S. Upton, vice president (retired senior vice president and treasurer, Whirlpool Corporation); Charles M. Odirizzi (executive vice president, Radio Corporation of America); Maynard M. Munch (special partner, Fulton, Reid & Company and president, The Maynard H. Munch Company); Frank M. Folsom (president, Radio Corporation of America); and Robert E. Brooker (vice president, Sears, Roebuck & Company). Photograph in the author's collection.

Seeger in the ratio of 1⅜ shares of such common stock and 3/16 of a share of such preferred stock for each outstanding share of Seeger common stock.

The Seeger Corporation brought to the merger discussions a treasury of paid-in capital of more than $28,500,000, the equivalent in 2005 inflation-adjusted dollars of $195,806,830. The issue at hand was how to protect the company's shareholder value residing in the financial reserve. If the funds were distributed by a special dividend to the stockholders, then both the corporation and the shareholders would be liable for tax payments. On the advice of the Seeger Company's brilliant legal counsel, Jack C. Foote, a partner of the St. Paul law firm of Doherty, Rumble, Butler, and Mitchell, the Seeger negotiators agreed to accept shares of convertible preferred stock in the Whirlpool-Seeger Corporation in lieu of a dividend. Under this arrangement, neither the Seeger Corporation nor the company's shareholders would suffer tax liabilities. In return, however, the Seeger representatives in the merger discussions forfeited the opportunity to receive an

Jack C. Foote. Photo courtesy of Mrs. Jack C. Foote.

increased number of shares of common stock sufficient to control a voting majority of stockholders in the merged corporation. Thus the decision to accept the preferred stock placed Seeger interests in a minority position on voting issues related to the governance of the merged corporation. It also foreclosed Seeger representatives from a prominent management role in the merged firm. RCA received, in conversion of the outstanding capital stock of the Delaware Appliance Corporation, twenty percent of the shares of Whirlpool-Seeger Common Stock issuable in connection with the merger.[15]

On the basis of shares of common stock of Whirlpool and Seeger outstanding on July 29, 1955, the Whirlpool common stockholders received an aggregate of 3,086,024 shares of common stock of Whirlpool-Seeger, the Seeger common stockholders received an aggregate of 1,548,229 shares of common stock and 211,122 shares of 4¼ percent cumulative convertible preferred stock of Whirlpool-Seeger, and RCA received 1,158,563 shares of common stock of the new company.

Whirlpool-Seeger would have an authorized capital stock consisting of 10,500,000 shares, of which 10,000,000 shares was common stock, par value $5 per share, and 500,000 shares was preferred stock, par value $80 per share, issuable in series.

Upon the merger becoming effective, Whirlpool-Seeger would have outstanding 5,792,816 shares of common stock and 211,122 shares of 4¼ % cumulative convertible preferred stock.[16]

In summary, the allocation of common stock in the merged Whirlpool-Seeger Corporation of September 15, 1955, broke out in the following manner:

RCA received 1,158,563 shares of common stock (20.0%)
Seeger received 1,548,229 shares of common stock (26.7%)
Whirlpool received 3,086,025 shares of common stock (53.3%).

Thus those individuals and other entities that owned stock in Whirlpool prior to the merger became the controlling interest in the newly formed Whirlpool-Seeger Corporation.

11

1955–1984: The Merged Corporation

THE TASK FOR THE THREE corporate founders of the new Whirlpool-Seeger Corporation was to make the merger work. Each participant had contributed vital ingredients to the new corporation, now the assignment was to build a complicated enterprise on that foundation.[1] In 1955 and 1956 nearly 90 percent of the consolidated sales revenues of the merger parties had come from appliances manufactured for just one customer—Sears. Consequently an immediate and high priority corporate goal for Whirlpool-Seeger was to develop an independent line of appliances bearing the *RCA Whirlpool* brand name. Between 1956 and 1968 Whirlpool had the right to add the *RCA* label to the appliances it manufactured. In retrospect this calculated marketing strategy paid immediate dividends because it aided the development of an independent line of appliance sales. By 1970 the sales account with Sears represented 64 percent of the company's revenue, a decline of 26 percent points from the 1956 figure. The loss of revenue from Sears was, however, more than offset by the surge in sales of the *RCA Whirlpool* line of appliances.[2]

Between 1955 and 1970 the Whirlpool-Seeger Corporation and its successor, the Whirlpool Corporation, consolidated its resources and expanded rapidly, introducing new and improved lines of appliances, increasing production and sales, and enhancing earnings for its shareholders. By any measurement, the three parties to the merger had attained their sales and revenue objectives. The detailed analysis in Figure 1 of the St. Paul Division's operations from 1956 through 1967 including expenses, return on investment, average number of employees, the number of the various products manufactured and shipped, and the dollars generated by the sales of the products reveals a productive and profitable business operation. The Whirlpool Corporation was on the path to becoming the world's largest manufacturer of home appliances.

1956 Annual Report Whirlpool-Seeger Corporation

1956 is the first calendar year of operations for Whirlpool-Seeger Corporation. The first major objective of the Corporation was reached in December 1955 with the presentation to our distributor organization of the first full line of RCA WHIRLPOOL major appliances. At the end of 1955 the products of Whirlpool-Seeger consisted of laundry, refrigerator and freezer lines for Sears, Roebuck and Co., and a line of laundry appliances bearing the RCA WHIRLPOOL brand name. At the end of 1956, the laundry, refrigerator and freezer lines for Sears had been greatly expanded from 3 major appliance lines totaling less than 50 different models to 12 different appliances approximating 150 basic models.

Walter G. Seeger,
Chairman of the Board of Directors

FIGURE 1

Major Trends at the St. Paul Division of the Whirlpool Corporation 1956–1968

($ 000 omitted)

	1956	1957	1958	1959	1960	1961	1962	1963	1964	1965	1966	1967	1968 PLAN
TOTAL DIVISION													
Product Engineering (Incl. rebilled)			$ 1,048	$ 1,322	$ 1,110	$ 1,123	$ 1,233	$ 1,256	$ 1,240	$ 1,345	$ 1,548	$ 1,555	$ 1,516
Development			467	291	520	462	350	412	433	400	412	309	240
Total Engineering			**$ 1,515**	**$ 1,613**	**$ 1,630**	**$ 1,585**	**$ 1,583**	**$ 1,668**	**$ 1,673**	**$ 1,745**	**$ 1,960**	**$ 1,864**	**$ 1,756**
Capital Expenditures	$ 596	$ 403	352	581	1,028	1,008	344	1,153	281	699	884	928	1,281
Tooling Expenditurees	635	1,209	2,098	860	676	550	765	1,619	723	520	935	702	1,096
Sales	**$ 47,321**	**$ 35,348**	**$ 51,262**	**$ 55,333**	**$ 55,118**	**$ 48,238**	**$ 45,034**	**$ 47,208**	**$ 52,006**	**$ 56,536**	**$ 62,308**	**$ 66,364**	**$ 72,245**
Earnings before taxes	6,258	3,847	3,992	3,002	4,638	2,579	2,773	3,126	3,185	4,831	6,331	4,485	5,580
% to Sales	13.2%	10.9%	7.8%	5.4%	8.4%	5.3%	6.2%	6.6%	6.1%	8.5%	10.2%	6.8%	7.7%
Investment at Gross	$ 20,929	$ 19,848	$ 22,217	$ 24,578	$ 22,828	$ 21,427	$ 21,775	$ 24,198	$ 23,694	$ 24,040	$ 26,516	$ 28,984	$ 31,000
Return on Investment	**29.9%**	**19.4%**	**18.0%**	**12.5%**	**20.3%**	**12.0%**	**12.7%**	**12.9%**	**13.4%**	**20.1%**	**23.9%**	**15.5%**	**18.0%**
Average No. Employees	2,035	1,504	1,921	2,149	1,949	1,602	1,552	1,544	1,539	1,633	1,894	1,942	1,965
Square Feet	690,513	690,513	690,513	702,255	720,227	769,671	769,671	732,263	720,280	721,330	723,988	728,682	728,682
Units Shipped	180,795	152,887	499,012	538,649	570,295	528,341	505,752	631,274	691,370	759,245	855,929	893,008	916,007
FREEZERS & REFRIGERATORS													
Units produced	166,495	144,690	175,213	189,559	193,435	159,283	165,858	146,937	138,972	139,306	144,176	155,051	175,301
Direct labor generated	$ 2,601	$ 2,110	$ 2,878	$ 3,031	$ 2,866	$ 2,078	$ 2,081	$ 1,904	$ 1,820	$ 1,801	$ 1,852	$ 2,056	$ 2,210
Units shipped	158,967	134,538	188,968	181,424	186,394	163,982	158,547	153,318	142,937	144,597	137,240	160,594	174,086
Sales	**$ 28,598**	**$ 22,949**	**$ 32,813**	**$ 31,301**	**$ 27,926**	**$ 23,355**	**$ 21,808**	**$ 20,231**	**$ 19,209**	**$ 19,317**	**$ 17,486**	**$ 20,402**	**$ 22,290**
Trading Income	3,867	2,321	3,051	1,545	1,867	1,388	816	338	(191)	1,432	2,357	1,635	1,450
% to Sales	13.5%	10.1%	9.3%	4.9%	6.7%	5.9%	3.7%	1.7%	(1.0%)	7.4%	13.1%	8.0%	6.5%
Investment at gross				$ 12,623	$ 11,952	$ 10,511	$ 10,360	$ 11,742	$ 10,184	$ 9,392	$ 9,183	$ 8,888	$ 10,212
Return on Investment				12.2%	15.6%	13.2%	7.9%	2.9%	(1.9%)	15.2%	25.7%	18.4%	14.2%

FIGURE 1 CONTINUED

	1956	1957	1958	1959	1960	1961	1962	1963	1964	1965	1966	1967	1968 PLAN
ICE MAKING EQUIPMENT													
United produced	12,187	12,728	8,998	16,149	13,492	12,672	15,132	19,311	18,953	22,322	24,689	23,431	27,759
Direct labor generated	$ 582	$ 487	$ 392	$ 633	$ 496	$ 491	$ 597	$ 773	$ 757	$ 957	$ 1,031	$ 981	$ 1,124
Units shipped	11,593	11,816	10,564	14,044	13,861	13,326	14,468	17,096	19,602	22,048	22,974	23,993	27,180
Sales	**$ 3,112**	**$ 3,218**	**$ 2,779**	**$ 3,528**	**$ 3,428**	**$ 3,283**	**$ 3,750**	**$ 5,073**	**$ 6,266**	**$ 7,048**	**$ 7,479**	**$ 8,030**	**$ 9,451**
Trading Income	414	443	400	340	316	327	203	228	379	130	342	105	473
% to Sales	13.3%	13.8%	14.4%	9.6%	9.2%	10.0%	5.4%	4.5%	6.0%	1.8%	4.6%	1.3%	5.0%
Investment at gross				$ 2,022	$ 1,630	$ 1,429	$ 1,912	$ 2,536	$ 2,916	$ 3,694	$ 4,763	$ 4,216	$ 4,526
Return on Investment				16.8%	19.5%	22.9%	10.7%	9.0%	13.0%	3.5%	9.6%	2.5%	10.5%
MILK COOLERS													
Units produced	6,892	4,052	2,918	3,843	3,963	3,877	2,990	2,432	2,487	2,237	1,931	2,506	2,393
Direct labor generated	$ 779	$ 438	$ 366	$ 471	$ 511	$ 486	$ 355	$ 294	$ 298	$ 299	$ 262	$ 344	$ 340
Units shipped	5,715	3,645	3,659	3,870	3,957	4,064	2,956	2,413	2,831	2,178	2,259	2,272	2,400
Sales	**$ 6,830**	**$ 4,852**	**$ 4,943**	**$ 5,345**	**$ 5,469**	**$ 5,578**	**$ 3,758**	**$ 3,108**	**$ 3,920**	**$ 3,273**	**$ 3,377**	**$ 3,384**	**$ 3,617**
Trading Income	577	513	557	504	637	612	236	203	554	508	595	464	419
% to Sales	8.4%	10.6%	11.3%	9.4%	11.7%	11.0%	6.3%	6.5%	14.1%	15.5%	17.6%	13.7%	11.6%
Investment at gross				$ 2,674	$ 3,057	$ 2,984	$ 2,956	$ 2,400	$ 1,962	$ 2,031	$ 1,778	$ 2,142	$ 2,104
Return on Investment				18.8%	20.8%	20.5%	8.0%	8.5%	28.2%	25.0%	33.5%	21.7%	19.9%
FLOOR CARE (Vacs. Only)													
Units Produced			221,326	328,611	310,423	289,152	286,688	407,667	463,940	491,566	578,762	634,846	627,400
Direct labor generated			$ 605	$ 854	$ 796	$ 679	$ 567	$ 670	$ 816	$ 932	$ 1,152	$ 1,498	$ 1,407
Units shipped			240,323	310,253	313,550	295,628	275,965	394,543	451,876	497,361	572,907	592,077	630,810
Sales			**$ 8,982**	**$ 11,044**	**$ 10,842**	**$ 9,654**	**$ 9,285**	**$ 11,858**	**$ 14,331**	**$ 16,841**	**$ 19,898**	**$ 21,122**	**$ 24,362**
Trading Income			55	84	304	86	733	1,307	1,385	1,338	1,474	1,404	1,552
% to Sales			.6%	.8%	2.8%	.9%	7.9%	11.0%	9.7%	7.9%	7.4%	6.6%	6.4%
Investment at gross				$ 4,440	$ 4,008	$ 3,934	$ 3,659	$ 4,066	$ 4,820	$ 5,340	$ 7,538	$ 8,805	$ 10,048
Return on Investment				1.9%	7.6%	2.2%	20.0%	32.1%	28.7%	25.0%	19.6%	15.9%	15.4%
KITCHEN (Dishwasher & Disposer)													
Units produced			4,728	29,477	57,952	45,683	58,042	64,007	77,435	88,833	130,919	110,414	74,940
Direct labor generated			$ 8	$ 196	$ 478	$ 321	$ 439	$ 446	$ 613	$ 627	$ 996	$ 866	$ 660
Units shipped			3,926	28,794	51,666	50,568	53,079	62,728	72,978	92,134	119,635	112,885	81,531
Sales			**$ 163**	**$ 3,009**	**$ 6,529**	**$ 5,484**	**$ 5,397**	**$ 5,859**	**$ 7,295**	**$ 8,982**	**$ 12,026**	**$ 11,847**	**$ 10,346**
Trading Income			(121)	(62)	561	(323)	(244)	28	(86)	345	619	164	517
% to Sales			(74.2%)	(2.0%)	8.6%	(5.9%)	(4.5%)	.5%	(1.2%)	3.9%	5.1%	1.4%	5.0%
Investment at gross				$ 1,790	$ 2,787	$ 2,226	$ 2,569	$ 3,120	$ 3,612	$ 3,369	$ 4,129	$ 4,634	$ 4,127
Return on Investment				(3.5%)	20.1%	(14.5%)	(9.5%)	.9%	(2.4%)	10.2%	15.0%	3.5%	12.5%

Unlike the other divisions of the Whirlpool-Seeger Corporation, the St. Paul plant was a complete manufacturing facility that provided such operational services as financial, planning, purchasing, engineering, and laboratories. From the mid-1940s, the facility had been producing a variety of appliances for home and commercial purposes that reflected the corporation's expanding line of products. Commercial icemakers were manufactured for several private-label companies including Frigidare, Scotsman, Mile High of Denver, I.M. Cornelius of Minneapolis, and ACME of New York (1946–82). Residential icemakers were produced for the in-house brands of Whirlpool and Kitchen Aid, for Sears, and for outside companies such as SubZero, General Electric, and Frigidare. The St. Paul plant manufactured chest and compact freezers, bulk-milk coolers (1954–72), front-loading and portable dishwashers (1956–1969), food disposals (1958–65), and trash mashers. In 1956 Whirlpool-Seeger acquired the Birtman Electric Company of Chicago. This purchase extended the Whirlpool product line and in 1957 brought to the Arcade Street factory the manufacture of canister, upright, and built-in vacuums bearing the *Whirlpool* brand name

Whirlpool, David Oreck, and Vacuum Cleaners

The contrasting experience of David Oreck with the Whirlpool Corporation during the 1960s and '70s casts another spotlight on an episode of puzzling and destructive corporate management. Born in Duluth, Minnesota, in the early 1960s, Oreck was highly successful selling Whirlpool appliances and RCA televisions and microwaves in New York City. In 1963 he established the Oreck Corporation. At the time Whirlpool was unable to make a success of its upright vacuum cleaners and Oreck believed that with a redesign of the machine, he could give it a new lease on life. Whirlpool subsequently gave him exclusive rights to market its vacuum cleaners throughout the United States and also gave him the right to redesign their machine if he chose to do so. Oreck went ahead with the redesign and then produced his prototype under the RCA-Whirlpool brand name.

The Arcade Street factory manufactured Oreck's redesigned vacuum cleaners. David and his brother, Marshall, were frequent visitors to the plant and were acquainted with Walter Seeger and John Holl. As shrewd entrepreneurs and merchandisers, the Orecks quickly made the new vacuum immensely popular and profitable. By 1967 a declining market share among the competitors in the vacuum cleaner market prompted John Platts, the president of Whirlpool, to inform Oreck that Sears felt the Oreck vacuum cleaners were "troublesome." That comment strongly implied that the Oreck machine had now become a strong competitor to whatever machine Sears was selling.

Soon thereafter frequent problems began to occur in the manufacturing of the Oreck machines at Whirlpool. For example, the drive belts that were installed in the Oreck machines suddenly developed faults and flaws that caused the vacuum to fail to operate properly. Despite frequent complaints from Oreck, these manufacturing defects were not corrected. Oreck, therefore, had no recourse but to file a lawsuit against Whirlpool in July 1972. In that suit, Oreck complained that Whirlpool, which was also manufacturing the *Kenmore* private-label vacuum cleaners for Sears, had conspired to terminate Oreck's company as the Whirlpool distributor. The suit also claimed that Whirlpool and Sears conspired to keep his company from doing business in Canada. Four years later, in July 1976 a federal court jury found that Whirlpool and Sears, Roebuck & Company had indeed committed conspiracy to restrain trade in violation of the Sherman Antitrust Act and awarded David Oreck, $2,250,000 in treble damages.*

**Oreck Corp. v. Whirlpool Corp.*, 639 F.2d 75 (2d Cir. 1980).

A photo from about 1968 displays the cordial business relations between Walter G. Seeger (center) and Whirlpool's Elisha Gray II (left) and John Plotz. Photograph in the author's collection.

(1956–83). The East Side plant also manufactured vacuums from about 1962 until 1972 for Duluth-native, supersalesman David Oreck.[3]

Nonetheless inevitable changes transformed the management of the former Seeger plants in St. Paul and in Evansville, Indiana. The Seeger Refrigerator Company's facilities were now part of a larger corporation. No longer was the St. Paul Division's leadership in the hands of those linked to the Seeger legacy. Rather, a new breed of corporate executive came on the scene to shape policies and strategies that corporate officers based at Whirlpool headquarters in St. Joseph, Michigan, had crafted and approved. Day-to-day responsibilities for factory operations passed to a group of managers who subsequently became Whirlpool executives, including Bit Holder, Bud Wambach, Charles Findlay, George Warderberg, Robert Bartley, and Donald Casey.

Although Walter Seeger served as chairman of the board of the new corporation until 1958 and as a member of the board of directors until 1968, he had no responsibility for day-to-day operations. The last president of the Seeger firm, John Seeger Holl, became vice president of the St. Paul Division, continuing in that post until 1961. A team assembled at St. Joseph, Michigan, dominated the management of the new corporation

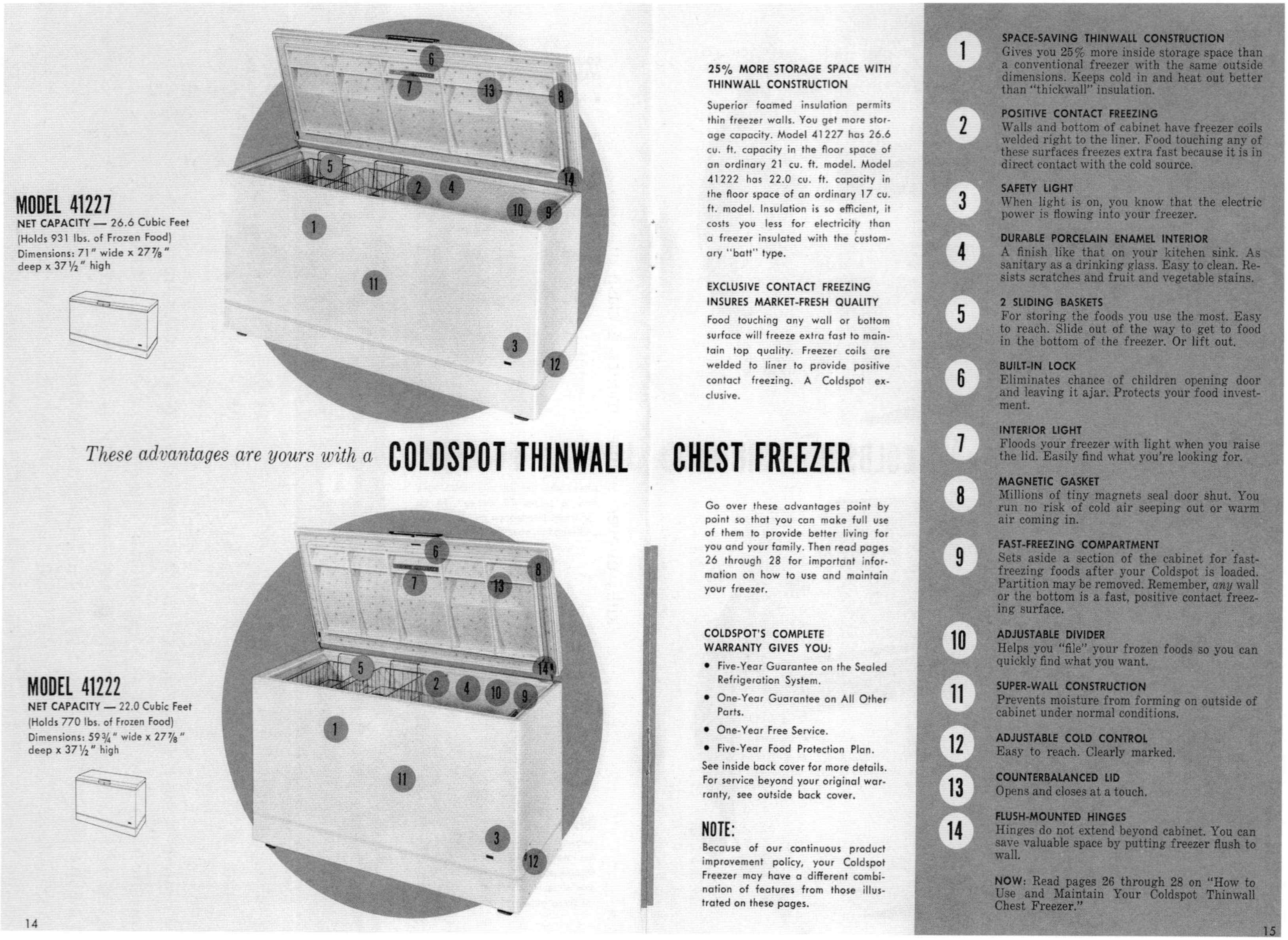

MODEL 41227
NET CAPACITY — 26.6 Cubic Feet
(Holds 931 lbs. of Frozen Food)
Dimensions: 71" wide x 27⅞" deep x 37½" high

These advantages are yours with a **COLDSPOT THINWALL CHEST FREEZER**

MODEL 41222
NET CAPACITY — 22.0 Cubic Feet
(Holds 770 lbs. of Frozen Food)
Dimensions: 59¾" wide x 27⅞" deep x 37½" high

14

25% MORE STORAGE SPACE WITH THINWALL CONSTRUCTION

Superior foamed insulation permits thin freezer walls. You get more storage capacity. Model 41227 has 26.6 cu. ft. capacity in the floor space of an ordinary 21 cu. ft. model. Model 41222 has 22.0 cu. ft. capacity in the floor space of an ordinary 17 cu. ft. model. Insulation is so efficient, it costs you less for electricity than a freezer insulated with the customary "batt" type.

EXCLUSIVE CONTACT FREEZING INSURES MARKET-FRESH QUALITY

Food touching any wall or bottom surface will freeze extra fast to maintain top quality. Freezer coils are welded to liner to provide positive contact freezing. A Coldspot exclusive.

Go over these advantages point by point so that you can make full use of them to provide better living for you and your family. Then read pages 26 through 28 for important information on how to use and maintain your freezer.

COLDSPOT'S COMPLETE WARRANTY GIVES YOU:

- Five-Year Guarantee on the Sealed Refrigeration System.
- One-Year Guarantee on All Other Parts.
- One-Year Free Service.
- Five-Year Food Protection Plan.

See inside back cover for more details. For service beyond your original warranty, see outside back cover.

NOTE:

Because of our continuous product improvement policy, your Coldspot Freezer may have a different combination of features from those illustrated on these pages.

1 **SPACE-SAVING THINWALL CONSTRUCTION**
Gives you 25% more inside storage space than a conventional freezer with the same outside dimensions. Keeps cold in and heat out better than "thickwall" insulation.

2 **POSITIVE CONTACT FREEZING**
Walls and bottom of cabinet have freezer coils welded right to the liner. Food touching any of these surfaces freezes extra fast because it is in direct contact with the cold source.

3 **SAFETY LIGHT**
When light is on, you know that the electric power is flowing into your freezer.

4 **DURABLE PORCELAIN ENAMEL INTERIOR**
A finish like that on your kitchen sink. As sanitary as a drinking glass. Easy to clean. Resists scratches and fruit and vegetable stains.

5 **2 SLIDING BASKETS**
For storing the foods you use the most. Easy to reach. Slide out of the way to get to food in the bottom of the freezer. Or lift out.

6 **BUILT-IN LOCK**
Eliminates chance of children opening door and leaving it ajar. Protects your food investment.

7 **INTERIOR LIGHT**
Floods your freezer with light when you raise the lid. Easily find what you're looking for.

8 **MAGNETIC GASKET**
Millions of tiny magnets seal door shut. You run no risk of cold air seeping out or warm air coming in.

9 **FAST-FREEZING COMPARTMENT**
Sets aside a section of the cabinet for fast-freezing foods after your Coldspot is loaded. Partition may be removed. Remember, *any* wall or the bottom is a fast, positive contact freezing surface.

10 **ADJUSTABLE DIVIDER**
Helps you "file" your frozen foods so you can quickly find what you want.

11 **SUPER-WALL CONSTRUCTION**
Prevents moisture from forming on outside of cabinet under normal conditions.

12 **ADJUSTABLE COLD CONTROL**
Easy to reach. Clearly marked.

13 **COUNTERBALANCED LID**
Opens and closes at a touch.

14 **FLUSH-MOUNTED HINGES**
Hinges do not extend beyond cabinet. You can save valuable space by putting freezer flush to wall.

NOW: Read pages 26 through 28 on "How to Use and Maintain Your Coldspot Thinwall Chest Freezer."

15

This page from a 1964 Sears freezer manual extols the advantages of its Coldspot *chest freezers, which were manufactured at the St. Paul Division of Whirlpool. From a Sears manual in the author's collection.*

and shoved aside the former Seeger leaders in St. Paul and Evansville. The bond between Walter Seeger and Elisha Gray remained strong and loyal during Gray's tenure as president and later chairman of the new corporation. Vision, purpose, character, and integrity were the hallmarks of Seeger and Gray; they liked and respected each other.

With Walter Seeger's resignation from the corporation's board in August 1968, Gray remarked that: "It marks the change of an era when a man who has been actively associated with a company for sixty-one years leaves its formal management structure. Walter Seeger started with the Seeger Refrigerator Company in 1907 and the strength and wisdom of his leadership helped make Whirlpool the great institution that it is today. I am sure I express the sentiments of all of our employees when I say that we all looked to Mr. Seeger with the greatest respect and affection."[4] For the merged corporation, the long shadows of Seeger and Gray had adroitly steered Whirlpool over the years between 1955 and 1968.

Seeger's resignation unmistakably signaled that many responsibilities of the company had been passed to a new generation of corporate leaders. In part, the transition was inevitable, a consequence of the passage of time for people and a business institution. A second generation of leadership in a family, a corporation, or a colony seldom maintains the vision, style, and character of the first generation of founders. This may be due to a changed perspective, differing talents and skills, uneven temperaments, a clash of personalities, or recognition that a new strategy was needed. Whirlpool's chief executive officers during the 1970s and most of the 1980s had not had the benefit of association with such captains of American finance, industry, and merchandizing as Robert E. Wood at Sears and David Sarnoff at RCA.

Nor had the new leaders been stamped with educations at such institutions as the Massachusetts Institute of Technology, the University of Chicago, or the University of Minnesota. In fact, between 1971 and 1987, the direction and leadership of the corporation was in large measure anchored on men whose experience had been garnered on the factory floor of the former Nineteen Hundred or Whirlpool Corporation, the manufacturer of laundry appliances.[5] With the advantage of hindsight, the eighteen years after 1968 was a quiet era for the reformed Whirlpool Corporation and did not mirror the dynamic vision of Seeger, Gray, and Sarnoff who had created the merged corporation in 1955. It was a lackluster period sandwiched between the dynamic leadership of chairmen Elisha Gray and David Whitwam.

Viewing the Whirlpool history after the merger with the luxury of a rear-view mirror, the passing of Walter Seeger gradually diminished the role of the St. Paul factory in the Whirlpool Corporation. Despite Seeger's

Hard Work, Education, and a Long Career at Seeger

Matthew C. Christensen is representative of the loyal, talented, and skilled employees associated with Seeger over many years. An Iowa native, he moved to Minneapolis in 1926 and held a number of manufacturing jobs before joining Seeger in January 1937. Initially his wage was 45¢ per hour. By February 1939 his monthly paycheck was $53.51, rising to $284.86 in 1946. Because Seeger was a union shop, Christensen was required to join the Refrigerator Workers Union. His 1941 Union Membership Card is illustrated here with the stamps indicating that he had paid his monthly dues.

In addition to his job, Christensen enrolled in mechanical drafting classes at the Dunwoody Institute in Minneapolis. While he was a student at Dunwoody, he may have met Robert S. Ahrens, the chief engineer at Seeger who when he was younger had been a student at Dunwoody in the 1920s and subsequently became a faculty member there. Christensen continue his education during the 1930s and 1940s by enrolling in evening classes at the University of Minnesota.

By 1961 Matthew Christensen's job title was Tool Engineer, Manufacturing Engineering Department and three years later he was a Welding Engineer in the Manufacturing Process Development Department. An engineer by inclination and training, Christensen was driven to exercise precision and quality in his tasks. In the 1960s the Whirlpool management submitted several of his designs for innovative improvements to their refrigeration products for patent review.

This photo from 1949 shows Matthew Christensen (right) working on a door assembly problem. From a Seeger News *in the author's collection.*

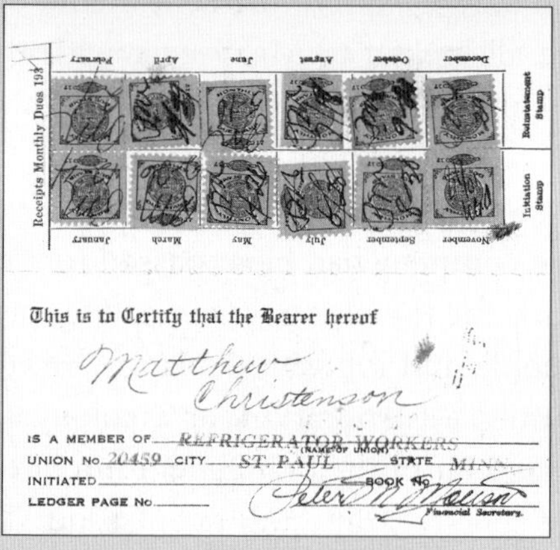

Receipts Monthly Dues 193

This is to Certify that the Bearer hereof

Matthew Christenson

IS A MEMBER OF REFRIGERATOR WORKERS

UNION No. 20459 CITY ST. PAUL STATE MINN

INITIATED BOOK No.

LEDGER PAGE No.

Financial Secretary

Stamps in Matthew Christensen's union membership book indicate he was paid up. Photo courtesy of the family of Matthew Christensen.

In 1970 Matthew Christensen received a large gift at the time of his retirement from Whirlpool's St. Paul Division. Photo courtesy of the family of Matthew Christensen.

long tradition of experience with talented, skilled, and loyal employees, and the distinguished record of the company's manufacture of a variety of residential and commercial appliances, the Seeger factory was virtually ignored by the highest corporate officers. The corps of St. Paul and Evansville employees in all likelihood held more patents on products the company manufactured than any other division of the new corporation. As Gene Gardner remarked in an interview for this book, "the St Paul plant was, unlike the other [Whirlpool] divisions, a complete manufacturing facility including financial, engineering, labs, planning, and purchasing, which was an ideal training ground for future executives. I know that this made it possible to develop various products, but I also believe that our ability to work

These Whirlpool employees are assembling chest freezers at the St. Paul plant in 1975. Sully Doroshaw photograph. Photo courtesy of the St. Paul Pioneer Press.

with each other to solve the tough problems that come up when developing new things had a lot to due with our success!"

Between 1955 and 1970 the St. Paul Division developed and manufactured several new products for Whirlpool. In a sense the plant was a microcosm in the production of appliances by the Whirlpool-Seeger and Whirlpool corporations. During these years the Seeger factory continued to make refrigerators and freezers, icemakers, and milk coolers, but it also introduced the production of vacuum cleaners, dishwashers, sink waste disposals, and compact freezers.

After Whirlpool-Seeger acquired the Birtman Company of Rock Island, Illinois, a manufacturer of a line of vacuum cleaners for Sears, production of the appliance was moved to St. Paul where the vacuum was completely redesigned and expanded to a full canister and upright line named *Kenmore* for Sears. As Figure 1 indicates, 221,326 units were manufactured in 1958 and 240,323 shipped to customers, By 1967 the factory was producing 634,846 vacuum cleaners and shipping 592,077.

Whirlpool launched a new appliance in 1958, the kitchen dishwasher. The

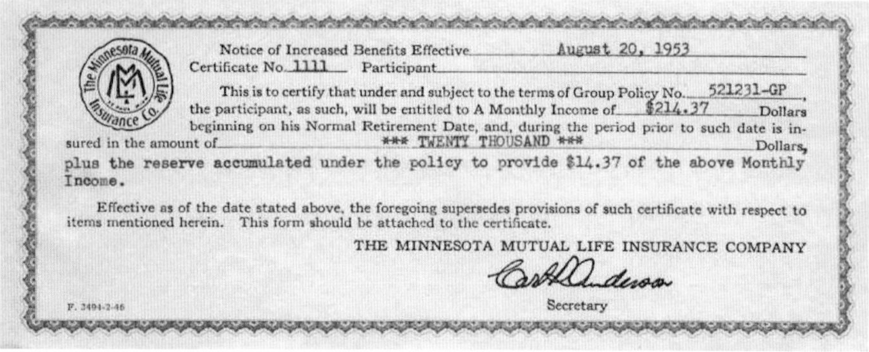

Notice of Increased Benefits Effective August 20, 1953
Certificate No. 1111 Participant

This is to certify that under and subject to the terms of Group Policy No. 521231-GP, the participant, as such, will be entitled to A Monthly Income of $214.37 Dollars beginning on his Normal Retirement Date, and, during the period prior to such date is insured in the amount of *** TWENTY THOUSAND *** Dollars, plus the reserve accumulated under the policy to provide $14.37 of the above Monthly Income.

Effective as of the date stated above, the foregoing supersedes provisions of such certificate with respect to items mentioned herein. This form should be attached to the certificate.

THE MINNESOTA MUTUAL LIFE INSURANCE COMPANY

Secretary

F. 3494-2-46

When Seeger and Whirlpool merged, the new corporation inherited all the employee benefits that the predecessor companies had in place in 1955. The Seeger Refrigerator Company brought into the merger retirement, disability, death, savings and other benefit plans for its workers, as illustrated here. From a 1953 benefits plan in the author's collection.

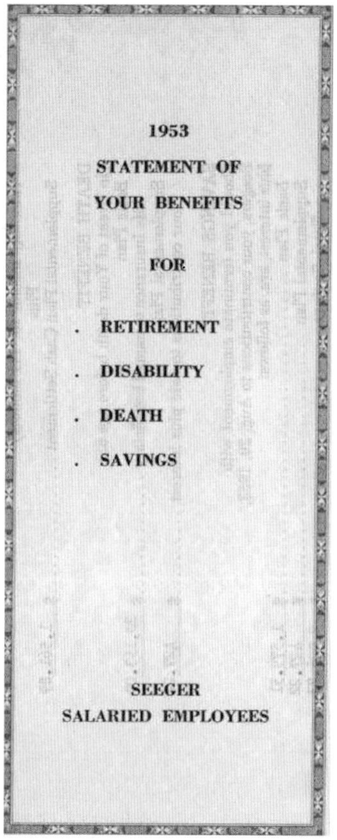

1953
STATEMENT OF
YOUR BENEFITS
FOR
. RETIREMENT
. DISABILITY
. DEATH
. SAVINGS

SEEGER
SALARIED EMPLOYEES

laboratory in Benton Harbor, sent the St. Paul Division a hand-built experimental machine that could not possibly be produced on an assembly line at anywhere near a reasonable price. The talented and skilled employees in St. Paul redesigned it, tooled it, tested it, and started building a line of portable and built-in dishwashers under the Whirlpool name. Inevitably there were startup problems and a number of St. Paul plant representatives made demonstrations around the country and into customer's homes to correct them. Only 4,728 units were manufactured in 1958, and 3,926 were shipped to customers. In 1967 production had increased to 110,414 units with 112,885 of them shipped. Soon afterward the volume had expanded to a point that a new plant was built in Ohio solely for manufacturing dishwashers.

Seeger began making commercial icemakers in the early 1950s. By 1956 the St. Paul plant had manufactured 12,187 units and shipped 11,593. A decade later the numbers were 23,431 and 23,993, respectively. A Model 50 icemaker was introduced later. Essentially a reduced-sized version of the large commercial models, it was to meet the growing demand for an icemaker to be used in private homes, bars, patios, and so forth.

Launched in 1954, Seeger built milk coolers for the Sunset Milk Cooler Corporation of St. Paul. Two years later 6,892 units were manufactured and 5,715 were shipped. By 1967, however, the production had declined to 2,506 units, with 2,272 shipped.

Today there is little acknowledgement of the role the Seeger Refrigerator Company played in the formation of Whirlpool-Seeger and the later

> **1957 Annual Report**
> **Whirlpool Corporation**
>
> **No mention was made in the report to the shareholders of the change of the Corporation's name on January 1, 1958, from the Whirlpool-Seeger Corporation to the Whirlpool Corporation.**

Whirlpool Corporation. This omission from the Whirlpool story suggests a willful and intentional corporate policy to cleanse and simplify the historical record, and it is unclear why that history should be so distorted. This attempt to obliterate the origins of the modern-day firm is at once misleading and a misrepresentation of the record, that's reminiscent of Topsy, that lovable and charming character in Harriet Beecher Stowe's classic novel, *Uncle Tom's Cabin*, who did not know her age, who declared that she never had a mother, and "never was born." Despite the leadership of Walter Seeger as the first chairman of the merged company, who played a major role in formulating corporate policy and procedures, served as the lead contact with David Sarnoff, forged financial links to such New York investment houses as J.P. Morgan and Company and the National City Bank, acted as a seasoned and highly respected executive in the field, and worked as a key public spokesman for the firm, the Whirlpool Corporation has not recognized his contribution. On the day of Walter Seeger's funeral in St. Paul, January 10, 1969, the production facility on Arcade Street paused for two minutes in his honor. According to the *Herald Press* of St. Joseph, Michigan, in January 1969 Whirlpool's chairman, Elisha Gray II, praised Walter Seeger in his remarks following Seeger's death and called him "an engineer of the American dream." Gray went on to say that Walter Seeger "was a giant of our industry and a friend and counselor to those who where fortunate to have known him. His wisdom and companionship will be missed by many." Elisha Gray's insightful reflections on Walter Seeger's business leadership are an eloquent testimonial to the warmth of their personal and professional association.

The Merged Corporation: 1968 to 1978

Between 1968 and 1977 Robert C. Upton, a Whirlpool group vice president who was headquartered in Benton Harbor, Michigan, supervised Whirlpool Corporation's St Paul Division. A third-generation descendant of the founder of the Upton Machine Company and a 1938 graduate of the University of Chicago, Upton was a principal party in the 1955 negotiations that led to the formation of the new Whirlpool-Seeger Corporation. His management and leadership style was akin to that of Elisha Gray and Walter Seeger, and he represented the last link to the men who had founded of Whirlpool-Seeger and its successor, the Whirlpool Corporation. After coming to St. Paul and touring the facilities in August 1968, Upton reported to the corporation's board "The St. Paul Division has the outstanding reputation within

Whirlpool to absorb in all its facets engineering, manufacturing, personnel, and so forth; [and put] new and small volume products into production quickly, efficiently, and economically." He observed that this practice made St. Paul much more of a job-shop operation than Whirlpool's other divisions. The other divisions within Whirlpool, Upton observed, typically operated on large-volume products, single-purpose manufacturing techniques, with the accompanying attention to smooth scheduling, long-range planning, and careful material control. He said that two products, floor care and the chest freezer, were the most important appliances produced in St. Paul since World War II. The chest freezer alone, he noted, "is a product which I'm sure is the largest and best operation in this tough industry and we are very proud of it."[6]

As for floor care, he noted, St. Paul made a complete line of such products for Sears. In addition, he said, the Arcade Street factory had been manufacturing dishwashers since 1958 and also was making bulk milk coolers for dairy farms, containers for fresh meat storage and handling, and was operating a profitable commercial icemaker business. Upton added that "the St. Paul Division, which has 728,682 square feet, another 250,000 square feet of leased warehouse space, and approximately 2,100 employees, has had a very successful and profitable history and we intend to carve greater successes in the future. I found the organization one which is extremely hardworking and enthusiastic."[7]

In January 1977, Upton announced his retirement from Whirlpool as of March 15, but he continued to serve on the company's board of directors. For nearly a decade, he had been a wise and supportive champion of the St. Paul Division. After his departure, the St. Paul factories on Arcade Street and in Cottage Grove were in decline. The escalating events that led to this decline have been partially described in a useful study by Wright Anderson Williams, an economics student at Macalester College.[8] The review of the facts that follows accurately reinforces the substance and interpretation found in his excellent analysis.

Ill Winds Swirl around St. Paul, 1978–1984

In retrospect, the seeds of dismantling and obliterating from memory the St. Paul operations of the former Seeger Refrigerator Company were sown possibly as early as March 1967. The first signal appeared in the company's weekly employee newssheet, *Inside Whirlpool News*. Two articles cast shadows on the St. Paul operations, indicating that the wages paid Seeger workers were higher than those that Whirlpool paid its employees in Ohio, Kentucky, and Arkansas. While the *News* noted that the St. Paul plant main-

Looking down Arcade Street in about 1980, this photo shows how Whirlpool's St. Paul Division plant straddled the street. City of St. Paul Planning and Economic Development photo courtesy of the Ramsey County Historical Society.

tained an excellent record for efficient productivity, its location was not close to necessary resources, and that the expenses of wages and medical costs in Minnesota were higher than at the other Whirlpool locations. The October 19, 1971, *News* lodged a similar criticism when it compared the high state taxes the company paid in Minnesota with those it paid in other states where Whirlpool had operations. While such economic criticisms may have been a strategy Whirlpool applied in labor union negotiations, they were also a factor in the decision to close the St. Paul facilities.

For the next five years after this comment on the high state taxes in Minnesota, the former Seeger operations ran more or less smoothly. In August 1976, Whirlpool sold one of its plants in Evansville, Indiana, to private investors. The factory's 600 workers produced evaporator and condenser coils for refrigeration units. Its closing was said to be a part of a program to reduce dependence on the Evansville facility and to transfer operations to the firm's other factories. Indeed, one refrigerator assembly line had been transferred to a plant in Fort Smith, Arkansas.[9]

In St. Paul in the mid-1970s, the Arcade Street plant recorded its highest employment ever with 2,600 employees working there before Whirlpool built a second plant in nearby Cottage Grove in Washington County in 1976. The Cottage Grove factory produced floor care operations, such as vacuum cleaners, while the Arcade Street facility continued to build freezers.

Unfortunately, by the 1980s foreign competition appeared to take its toll on Whirlpool's corporate profits and demand for American-made freezers, particularly chest freezers, and also vacuum cleaners declined.

After a mere seven years in operation, the Cottage Grove plant was closed and, soon after that, so was the former flagship Seeger factory on Arcade Street. The position of Whirlpool officials was that the corporation had more manufacturing facilities than it needed, that the division located in Danville, Kentucky, was an underused, modern, one-story plant, and that the Evansville factory could absorb the freezer production.

Since the 1930s Refrigerator Workers Union of Teamsters Local 827 had represented Seeger employees, and it had initiated only a few strikes. A thirteen-week strike in 1955 focused on two issues, worker dissatisfaction with wages and working conditions. Ironically, the company's settlement was better than the union's proposal. A five-weeks-long strike in 1974 was concerned with complex issues even though Whirlpool may have provided decent wages and benefits. The relationship between the union and Whirlpool may have appeared peaceful, but there seems to have been some attempts by Whirlpool in the late 1970s and early 1980s to cut the social contract and increase profits. In 1979, for instance, seven employees with the International Union of Operating Engineers went on strike for almost two weeks after contract negotiations broke down. Whirlpool and this union, however, eventually reached a settlement that included pay raises and fringe benefits. Later, in mid-August 1981, a brief strike by the Millwrights and Machinery Erectors Local 548 lasted two days. At issue was contract language and wages. In the case of this strike, worker loyalty to Whirlpool appeared high, since most of the members of Refrigerator Workers Union of Teamsters Local 827, the largest union in the plant, crossed the picket lines and, thus, the plant continued to operate. Most workers did not wish to risk lost wages by supporting a small number of dissatisfied employees.

Early warning signs of an impending shutdown of the St. Paul factory can be seen in the layoffs that took place in August 1982. Due to seasonal demand and a growing national recession, 300 production workers were laid off at that time. According to the company spokeswoman, freezers were not selling as well and, therefore, layoffs were required.[10] Yet Robert Bartley, the division vice president in charge of Whirlpool's St. Paul operations, remarked to the employees in an issue of *Inside Whirlpool News* that as he looked back over 1982:

> I can't help but be proud of the St. Paul Division's performance in weathering the difficult economic times we have experienced. . . . Our list of accomplishments includes improvements in product quality, redesigned central vacuum, redesigned chest freezer, new powder coat paint system, new polystyrene presses, new powermate motor

> and integration, and redesigned mid-line powermate through employees' participation. The challenge of 1983 will be to continue our division's spirit of innovation and working side-by-side to make this coming year a gainful one. I know the people of the St. Paul Division will continue providing the excellent support we need to attain these objectives.

Bartley's experience and sentiments seem to have been out-of-step with the thinking of Whirlpool brass at headquarters. Of course they may have kept their strategies secret as they unfolded the legal ploys that would shut down the St. Paul factories.

Another indication of how Whirlpool's top management altered, and may have undermined, production operations at the St. Paul factory occurred in the early 1980s. Perhaps as early as the late 1920s, but certainly by the mid-1930s, Frigidaire had ordered from Seeger commercial display cases that carried the Frigidaire label. Frigidaire sold the display cases to grocery stores and meat markets. For reasons that are not clear, Whirlpool's chairman Jack Sparks initiated what many at the time considered to be an ill-advised and ill-considered move by abruptly terminating the long-standing production relationship between the St. Paul Division and Frigidaire for these products, which had been one of Frigidaire's hallmark associations with the Seeger legacy.

In the early 1980s, Whirlpool claimed to have conducted two studies of the Cottage Grove and the Arcade Street plants. Consequently, management decided to move all production to Kentucky and Indiana. Despite the past achievements and continuing productivity of the Minnesota facilities across several decades, Whirlpool publicly declared in the spring of 1983 that it continued to seek solutions to poor profitability. It also ran a study to examine the need for manufacturing capacity and demand for products produced at the Cottage Grove plant.[11] At the same time, company officials insisted that there were no plans to shutdown the St. Paul Division, and that the study was being conducted merely "to structure the St. Paul Division in order to remain competitive in the challenging appliance industry," according to Bartley.[12] One employee, Ron Hanson, however, remembers that the company gave some workers the impression that Whirlpool planned to remain in the Twin Cities for a long time. Even so Whirlpool's spokesperson declared that shifting the floor care operations to the Danville (Kentucky) Division would be more cost-effective, and cited lower transportation costs and property taxes as additional reasons for moving the Cottage Grove operations.[13]

The decision to move floor care operations out of Minnesota at the time can be compared to the national trend of capital mobility. Workers at the Danville Division received less pay and fewer benefits than those in St. Paul. Kentucky, at that time, was a Right-to-Work state, and thus the Whirlpool plant there was non-union. Whirlpool, like many other companies in the 1980s, closed the St. Paul Division and transferred production to Kentucky and Indiana because those sister states offered a "good business climate." In other words, tax breaks were abundant, labor and transportation were cheap, and the savings supposedly were extraordinary. The average wage in Danville was $5 to $6 dollars an hour, compared with about $9.50 in St. Paul. Whirlpool predicted that moving the floor care operations to Kentucky would save an estimated $8.5 million in the next five years.[14]

Whirlpool essentially managed to avoid union protests by claiming that the Cottage Grove move was not based on labor costs, that wage concessions were not an issue. Floor care workers, however, were not allowed to transfer to Danville. Since workers were unable to use wage concessions as a bargaining tool, 250 jobs were lost in Cottage Grove and 150 jobs were lost at the Arcade Street factory. The Arcade plant continued operations with 1,000 workers.

The opinions of Cottage Grove workers at this time naturally revealed animosity. Most felt betrayed when their jobs were eliminated. According to Cottage Grove worker Judy Davidson: "At contract time they've always said this plant is a moneymaker. I feel like I've been stabbed in the back." Others felt that union leaders didn't tell them of these layoffs far enough in advance. Dan Bogstrom, a worker with enough seniority to transfer from Cottage Grove to the Arcade Street plant, thought the Cottage Grove closing was a wise decision. "They're only paying $18,000 a year in taxes on the Danville Plant. They're paying $300,000 here."[15]

On April 19, 1984, Whirlpool announced it was considering closing the Arcade Street factory and would undertake still another "study" over the next two months to examine the possibility of consolidating refrigerator and freezer production at other company plants.[16]

Surprisingly, one week later on April 26, 1984, the company announced that the Arcade Street factory, identified with the Seeger family and their refrigerator company since 1910, would close at the end of the year.[17] Despite the intervention of elected officials and union leaders in Minnesota, the decision stood. The conveyor belt ran for its last day on Friday, December 28, 1984; every twenty-two to thirty-three seconds a finished freezer rolled off the assembly line.[18] According to Whirlpool spokesperson Zora Schultz, shutdown was a decision based on a need for greater efficiency. "We had excess capacity throughout the corporation. There was room in Evansville to put the product. We needed less space to manufacture chest freezers because of the declining market and I guess we don't really foresee that turning around. The most effective alternative for the corporation was to move [the freezers and icemaker production] to Evansville."[19]

In a special edition of Whirlpool's daily employee newsletter, *Twelve O'Clock High*, of April 26, 1984, division vice president Donald Casey explained the factors that led to the shutdown.

> The St. Paul Division Product Analysis has been completed and the tentative decision is to relocate freezer and icemaker production to the Evansville Division in Indiana. Under this plan, it appears that the movement of these products will be completed late this year [1984] or early in 1985 . . . in the final analysis, our products will be more cost effective if consolidated [with] another division. In addition, the market for chest and compact freezers has decreased during recent years and is not expected to significantly improve. Of course, this makes competition to sell these products even more intense.[20]

In addition to its excess capacity, Casey also pointed out that the Arcade Plant was inefficient because it was very old and had many floors. Today's industrial plants, he explained, require better technology to increase the rate of production.

Negotiations for closing the Arcade Street plant reveal that Whirlpool's management would not allow its assembly workers to transfer to factories at other locations nor would they offer, according to the workers themselves, an acceptable severance package, acceptable retraining, or retirement benefits for those near fifty-five years of age. In addition, lawsuits were filed seeking compensation for a lack of safety measures at the Arcade Street plant. The Whirlpool closing in East St. Paul disastrously affected workers, their families, and the surrounding community.

In fact, plants elsewhere in Minnesota and the Upper Midwest were being closed down by their corporate owners in an effort to compete more effectively and to take advantage of cheaper labor costs in other parts of the country. At the time of the Arcade Street plant closing, workers felt that Whirlpool had sabotaged the St. Paul operation and had lied to them and cheated them. Some felt that they were betrayed. Robert Gruhlke, a thirty-one-year employee, was much disturbed at the time: "It stinks. They tell us nothing except lies for two years and then this." Marge Butler, a sixteen-year-employee agreed: "I'm mad. Now we'll all be out on the streets looking for jobs. You'd think after all that time you'd get more thanks and consideration."[21] Twelve years later, Ron Hanson still felt a sense of betrayal:

> We were down to, when the floor care moved, the lowest seniority was thirteen years. I think everybody felt betrayed. The way they were talking about how they were going to build down in Cottage Grove or they started to build and they didn't complete it. You know? A lot of people bought houses down there in Cottage Grove. They would have been closer to work.

For many workers, the Seeger plant was not just a place of employment; it was also a place of social interaction. According to Bob Schwan, Ron Hanson, and Clinton Conrad, interdepartmental basketball and softball provided an opportunity for social activity, as did the Seeger Mens Club. The employees were reminded that gold watches had been given to those workers who were employed with Seeger or Whirlpool for a long time. In addition, Seeger and Whirlpool offered college scholarships and Christmas toys to the children of employees. All of these were part of the benefits of a good social contract common for many of the larger United States employers in the post-World War II years. When the plant closed down, some employees might have felt as though an old friend was stabbing them in the back. Their bleak suspicions were correct.

Pleas by George Latimer, mayor of St. Paul, Rudy Perpich, governor of Minnesota, for Whirlpool executives to consider alternatives to shutting the plant were hopeless. Whirlpool's management was unshakeable, and disquieting new evidence suggests that the truthfulness and integrity of the corporate officials should be revisited.

In the midst of the closing, Whirlpool management offered a severance package to workers, which most considered mediocre. Costing more than $5 million, the package included early retirement benefits for those over age fifty-five, and job search assistance for all employees. In addition, workers were offered $20 for each month they had been with the company and each worker received an average of $5,280—one-third the amount Whirlpool would save in pretax earnings by moving to Kentucky and Indiana. The company study said that by leaving, Whirlpool could save an estimated $14.7 million.[22]

The union and the workers felt that this was a poor severance package. Unfortunately, they had no negotiating

Voices *from the* **Plant Floor** *at* **Seeger**

St. Paul and Its Changing Economy

"The closing of industrial plants has been going on for many years. We know that Bohn Refrigerator went. American Hoist, Koppers, Coke—all left a big hole in the neighborhoods where they were situated [when they closed or moved]. The railroad shops are all gone now, and the homes of the workers have new owners. Every coal company in town is gone, and most creameries and milk companies are gone, along with the workers. Ward's, Golden Rule, Schoch's, Emporium, John Wood, several plating companies, Midway, Bethesda, and St. John's hospitals are gone too. Every one left many persons unemployed and impacted the neighborhoods."

Howard Pittlekow

power. Ron Hanson felt helpless. "You didn't really negotiate with the company 'cause they didn't have to give you nothing. You know, they can just say 'We're closin.' Good luck. Good-bye.'"

By denying workers a portion of their retirement pensions, Whirlpool clearly violated the social wage. Whirlpool workers seriously believed that if they worked hard and put in enough years at the company, then they would be rewarded in their old age with a pension. As is widely acknowledged, this is no longer typical for the American workforce and most newly created jobs do not offer these benefits. Once again Whirlpool followed the national trend and its workers suffered economically.

Many of those who lost retirement benefits as a result of the mid-term closing as well as those in their middle ages had difficulty finding new jobs. Income loss and the psychological strain of being unemployed for a long time took its toll on the workers' self-esteem.

Voices *from the* Plant Floor *at* Seeger

Fallout from Closing the Seeger Plant

"There was a group of people who worked at the plant during its last three or four years who took the hit when the plant closed. Those many fine black workers who never had a good job in their lives, finally realized that they could do a good job, and get good pay, and were looked up to by their family and friends. This was important to them, and the experienced they gained while working for Whirlpool-Seeger probably changed the world in which they lived."

Howard Pittlekow

The lack of financial and moral support from the company during the closing and the subsequent job losses themselves had a direct impact on workers and their families. The bitterness some felt resulted in marital problems and deaths. According to Ron Hanson,

> Some of the [workers] got hit pretty hard—mentally and financially destroyed. Loss of houses; they couldn't make their payments. Even though they went back to work, they couldn't keep up the payments they were making. And there were a couple of suicides you could relate to the closing. A lot of divorces.

These experiences reveal the great importance employment with Seeger-Whirlpool held for many people. Bob Schwan:

> I know one of them [a former employee] who committed suicide. Slashed his wrists. I talked to his widow and she said, "He lived for and died for Whirlpool." He'd come to work and he was one of the stewards there for years and he took his job so serious, you know. And you can't joke about it, you know. If somebody'd goof off or something, boy he's right there to say, "Hey, you better straighten out, you know." And there were some who got divorced. A good friend of mine—him and his wife divorced right after that. He was a foreman for a while and just before the plant closed they took him

> off for no reason of his own, but they were just cuttin' back. So he lost out on a lot of benefits and his wife just said she couldn't live with him. He was just so bitter about it—everything and anything and they got a divorce. I think there were a lot of cases like that. You can't adjust to change in your way of living.

The Arcade plant closing had incredibly adverse effects on family relationships.

According to those former employees who were interviewed, production levels at the plants in Kentucky and Indiana never reached those in St. Paul. The lower production volumes were a consequence, perhaps, of less worker pride and loyalty at the other Whirlpool locations. Because so many workers felt a deep sense of pride in Whirlpool, this outcome, nevertheless, is disheartening. Even so, production was high enough to reveal how seriously Arcade people regarded their work, until the very end. Ken Erickson commented:

> Rumor had it that the Kentucky plant didn't work out very well. That their production, and this is just a comparison—not real numbers—here they'd be making 500 refrigerators a day. Down there they couldn't get a hundred out of the workers down there. Okay? You had skilled workers, but people up here really cared about their jobs and wanted to do a good job. I think that's why everyone was so angry.

Another incident showing great pride in their work was an announcement in the fall of 1984 of record-breaking production for compact and flash defrost freezers. Long-time Seeger employee Bob Schwan said that, throughout the years, Refrigerator Workers Union of Teamsters Local 827 kept operations at Arcade well organized because the union acted as a good middleman between the workers and the company. Lacking this buffer and company loyalty, production perhaps lagged in Kentucky and Indiana. One wonders why Whirlpool did not consider these factors as important in their decision.

On that last day of production, about 625 workers were operating the company's production line on Arcade Street. Their average age was forty-seven, and they had worked for Whirlpool and Seeger for an average of twenty-two years. In fact, 345 of those 625 had worked for Whirlpool or its predecessor firms, Whirlpool-Seeger, and Seeger for more than twenty years.[23] The workers were understandably angry, disappointed, and sad at the plant's closing. The factory had been a strong social and economic presence on the East Side and in St. Paul for more than seven decades. For years their wives had prepared and packed their lunch boxes, steady salaries had paid mortgages, medical bills, and the college tuitions of children. In return the employees gave unstinting loyalty to the Seeger Company and its successors, an association that brought accolades for the manufacturing accomplishments of the St. Paul plant's operations.

One employee noted that, "Like Seeger before it, Whirlpool was considered a source of secure employment and decent wages on the East Side. It has been around for eighty years. If you didn't finish high school or didn't go off to college, you worked at Whirlpool. Some families are on the third generation there."[24] Employee Kathy Youngquist said, "If there is one thing the workers at Whirlpool will want to remember, it's the way they worked together, everybody is like a brother and sister, or mother and father, to each other."[25] Kathy Zieman, another former employee, said "Whirlpool was a good place to work for a lot of years."[26] After high school, Jim Shaw worked at the Arcade Street plant for thirty-seven years. At a five-year reunion picnic the former employees held on July 22, 1989, at Battle Creek Park, Shaw looked back: "You miss them a whole lot. No job will ever be quite the same."[27] Dick Hagberg, has remarked to the author that "I'm happy that I could be a part of that great organization for thirty-three-plus years. When you see what was accomplished, it feels good to know that you had a part." Gene Gardner, manager of Project Planning at the plant, commented, "Thanks for all you are doing to bring back memories from our working days." While Howard Pittlekow, a keeper of many stories of his employment at Seeger, Whirlpool-Seeger, and the Whirlpool Corporation, noted, "Thanks for listening—the buildings are all gone, but the memories are there."

A local newspaper photographer was on the scene at 850 Arcade Street when Whirlpool's St. Paul Division closed its doors for the last time on December 23, 1984. Photo courtesy of the St. Paul Pioneer Press.

Today the Whirlpool Corporation is the world's leading manufacturer of major home appliances.[28] The company has plants in fifteen countries and markets products under more than a dozen major brand names in more than 170 countries. It is a global corporation poised to serve an ever-expanding market in the twenty-first century. Its shadow in St. Paul, however, grows dimmer with each passing year; yet the legacy of John A. Seeger, Walter G. Seeger, Gustave R. Seeger, and John S. Holl is as vital and vigorous today as it was in 1910, 1920, 1950 or 1970.

While the Seegers lived and worked in the city, they contributed to the fabric of the community. In the late 1920s and early '30s John A. Seeger served as chairman of the United Improvement Council, a group of citizens concerned about revitalizing downtown St. Paul. He was as a member of the St. Paul Library Board and of the commission for the construction of the Civic Auditorium. Following in his father's footsteps, Walter Seeger served as president of the St. Paul Association of Commerce; president of the St. Paul Baseball Club (the original St. Paul Saints team); as a member of the board of directors of several St. Paul and Minneapolis corporations; and as a trustee of the Shriners Hospitals for Crippled Children throughout North America. John Seeger Holl, the last family member to serve as president of Seeger Refrigerator Company and who for more than twenty years was an officer of the Whirlpool Corporation, continued the family's tradition of service to community institutions. His final act of philanthropy enriched such institutions as Macalester College, the Presbyterian Homes, the Saint Paul Foundation, and Carleton College.

The bold initiatives of John A. Seeger, the son of German immigrants and a former apprentice cabinetmaker, who moved intrepidly into manufacturing iceboxes that created a vital business organization for St. Paul and Ramsey County. His far-sighted adaptation of the electric refrigerator, and the vision of his sons and grandson positioned the corporation as a powerful participant in an emerging world of global interdependence and helped make the Whirlpool Corporation one of the nation's largest companies.

The contributions of the Seeger family members' themselves to the life of the St. Paul community have been no less valuable than the strength and productivity of their company. Through the years the Seeger Refrigerator Company employed thousands of people as tens of thousands of home appliances rolled off the assembly lines for households and commercial establishments near and far. Despite all the economic changes and upheavals of the 1980s, this Seeger legacy is undiminished and deserves to be remembered.

12

Summary

THE HISTORY OF THE Seeger Refrigerator Company is a bittersweet chronicle in St. Paul's long history. It is at once an engaging and fascinating account of the extraordinary achievements of three generations of the family's management and leadership, an American success story. A business and industrial accomplishment aided by the contributions of thousands of loyal, creative, and productive colleagues. I intentionally refrain from applying the terms employees or workers to the ranks of persons who worked at the Arcade Street plant as they were not perceived by the management in that way. Seeger managers recognized them as unique and important team members—partners in the enterprise of manufacturing refrigerators, freezers, and other quality home appliances. Yet the historical account is marked with ironies and contrasts, of periods of unspoken trust and distressing uncertainties, particularly during the last thirty years of its presence on the East Side.

For example, as a historian, I have examined the founding and development of a complex company between 1902 and 1955. My research has marveled at the manner in which the company and its staff applied through the years industrial and technological innovations to manufacture its products. In contrast the years following the merger with Whirlpool and RCA in 1955 were indelibly marked in a starkly different manner. Whether intentionally or not, seeds of suspicion were sown and shadows of betrayal appeared to stalk and grip the imaginations and opinions of longtime employees as they encountered puzzling and unfamiliar corporate management, policies, and controversy.

Walter Seeger was a key participant and architect of the merged Whirlpool-Seeger Corporation in 1955. He served as its chairman until

Voices *from the* **Plant Floor** *at* **Seeger**

The Importance of the Seeger Contribution

"I did not fully realize the importance of the Seeger contribution to the merger with Whirlpool and the Appliance Division of RCA until after [my] retirement. I will never understand why Whirlpool top management chose to wipe out any reference to the Seeger Company in the "official" history of the corporation. Without Seeger, there very well could not have been a merger. When we [former Seeger employees] organized our W Club, we were told not to use the Whirlpool name and they never offered any help or even acknowledged that we existed."

Gene Gardner

1958 and was a member of the board of directors for another decade. Yet the pivotal legacy of the Seeger Company in the historical account of the Whirlpool Corporation has been puzzlingly ignored. Inexplicably Walter Seeger's service is not appropriately recognized in the meeting room of the board of directors at Whirlpool's corporate headquarters in Benton Harbor, Michigan. His photograph is mysteriously absent from the gallery of men who through the decades have served as chairmen of such predecessor firms as the Upton Washing Machine Company, the Nineteen Hundred Corporation, and the post-1958 Whirlpool Corporation. Apparently the Whirlpool-Seeger Corporation did not exist because there is no public acknowledgement provided of the role of that corporation as the immediate predecessor of today's Whirlpool Company. The story of the corporation's experience as it stands is inaccurate and incomplete and a misrepresentation of a leader who significantly aided the launching of its first and formative years.

A further example of the unevenness of the Whirlpool Corporation's design to publicly acknowledge its origins and legacy occurred around 1976. At that time corporate officials devised a Twentieth Anniversary Award for presentation to twenty-four Whirlpool distributors, presumably in recognition of their association with the company since 1956. Ironically, the award was not symbolized by a simple letter W as a reference to the company's name, or by some other appropriate characterization, but by a trophy-like representation of a pre-World War II-vintage wringer-washing machine. The many other products manufactured by Whirlpool at the time were not noted on the award, a gesture that symbolically carried the message that the company was essentially a laundry appliance maker. Yet, despite the facts recited annually in the corporation's published annual reports to shareholders since the merger with the Seeger Company in 1955, it had generated about forty percent of its annual revenue from the sale of refrigeration products.

In addition a further distortion of the Whirlpool story occurred at the Evansville division of the company in 1976. The occasion was a ceremony celebrating twenty-years of the company's activity in the community. No mention was made of the fact that the facility had been a part of the Seeger Refrigerator Company since 1945, nor was it recounted that Seeger had had a long-standing association with the site's predecessor firm, the Sunbeam Electric Manufacturing Company. This association had begun in 1928 with the initial planning, development, and manufacturing of the *Coldspot* refrigerator for Sears, Roebuck and Company.[1]

A later event, described by a longtime employee of Whirlpool-Seeger and the Whirlpool Corporation, noted that in the 1980s, the company produced a traveling exhibition highlighting all the U.S. divisions of the company, thereby portraying and celebrating the Whirlpool Corporation's 75th anniversary. Incidentally, it was an event that underscored again the misrepresentations of the origins of the word Whirlpool and the company's diverse roots. For some reason there was no mention in this exhibit of the pivotal contributions of the Seeger Refrigerator Company or the facilities that operated in St. Paul for nearly ninety years. Such an exhibit, with a limited scope, not only does not accurately recount the company's heritage, it also missed an opportunity for broadcasting and interpreting a fascinating and dynamic American business story.

The massive and catastrophic destruction of the Seeger historical records when the plant was closed in December 1984 was a further and nearly definitive effort to obliterate the Seeger Refrigerator Company's heritage and legacy to the Whirlpool Corporation and to the community, state, and nation it served . While many records of the Whirlpool Corporation's operations in St. Paul may have been substantively transferred to the corporate headquarters in Benton Harbor, Michigan, the records of the Seeger Company were not. Former employees of the company have reported to me that they witnessed the disposal of enormous quantities of manuscript and printed materials relating to Seeger's history between 1902 and 1955. Several witnesses to the destruction and disposal of the records have stated that the materials destroyed included ledger and journal books; blueprints of products and parts manufactured at the plant; catalogs of products manufactured by the company for more than fifty years; correspondence files; photographs of the factories and operations in St. Paul and Evansville, Indiana; and other materials relating to the company's history and operations.

Voices *from the* **Plant Floor** *at* **Seeger**

Grateful

"I feel privileged and grateful that I could spend all my working years at the Seeger Refrigerator Company plant."

Richard Hagberg

Also lost were half-a-century of personnel records of the thousands of employees; the records of Seeger's clients; measurements and descriptions of the many refrigeration products manufactured by the company; data relating to personnel and production costs, numbers of units manufactured each year; information and contracts relating to wartime production; details and statistics relating to the sales organization and annual sales performance; and the yearly financial position of the company. The Whirlpool Corporation in one stroke wiped out an important aspect of the history of the refrigeration industry in the United States and the role of the Seeger family and the thousands of persons associated with the company in St. Paul, Minnesota, for nearly a century.

Recognizing the intrinsic value of the materials, a handful of loyal and long-serving employees gathered and preserved through the years a small

Demolition began in May 1987 on Whirlpool's former St. Paul Division factory on Arcade Street as part of an $8 million redevelopment project to be known as Seeger Square. Spence Hollstadt photograph. Photo courtesy of the St. Paul Pioneer Press.

remnant of the historical records. They have generously shared and passed to the author many of the items for safekeeping in perpetuity. These materials have been indispensable in researching and writing this book. It is inexplicable why Whirlpool Corporation officials did not appropriately offer the records of the Seeger Company for preservation and use by students to such institutions as the Minnesota Historical Society, the University of Minnesota Library, the Harvard Business School Library, or the Library of Congress. The destruction of the materials may be arguably characterized as a willful and deliberate act of cultural and historical negligence, especially as the family and their interests had a long and distinguished business and public presence in Minnesota since 1848.

Finally, we must remember that Whirlpool corporate officials decided under unusual circumstances to summarily terminate the last vestige of the Seeger Refrigerator Company's operations in St. Paul, an action in marked contrast with the Seeger reputation. In light of the murky details that surround the decision to close the St. Paul plant, the situation suggests that the judgment was a curious, unenlightened, and impulsive act. This conclusion

As demolition on the former Seeger site continued, soon all that remained was a pile of rubble that was being hauled away. Scott Takushi photograph. Photo courtesy of the St. Paul Pioneer Press.

is underscored by the fact that no copy of "the studies" has survived, if, indeed, any were ever written. The minutes of the Whirlpool Corporation's board of directors do not note that a "study" or "studies" were underway to determine if the plants should be closed, a discussion and vote to close the facilities, or that the factories had been closed. The circumstances surrounding this decision casts unmistakable doubt on the good faith and the integrity of Whirlpool's corporate leadership. Perhaps the board of directors was not an "independent" body at the time responsibly serving the shareholders' best interests but merely a rubber stamp for the management of the day. In any event the act does not represent responsible, prudent, and ethical corporate leadership. It is a management decision worthy of a systematic case study by future business students because it raises many significant questions reflecting on the integrity and ethical conduct of the corporation's leadership and the absence of any informed participation by the board of directors.

The 1984 decision by Whirlpool Corporation executives to close the St. Paul plants was a departure from the style of the company's past experience. For nearly seventy years the Seeger Company and its leaders had displayed an extraordinary sense of corporate responsibility to its employees and to the community. Its heritage and legacy to the community were jobs and high-quality products. It was common for two or three generations of a St. Paul family to have been associated with the company. Summer jobs were a financial lifeline for students seeking to earn money for their college expenses and such community activities were persistent and immeasurable contributions by all employees including the United Way, Blood Drives, the school board, state legislature, churches, schools, colleges, and hospitals. In return Seeger earned the respect and loyalty of its associates, reflected to this day by the dwindling band of employees who proudly identify themselves as Seeger alumni.

For a company without a coherent historical legacy, it is difficult to understand why Whirlpool would seek to obliterate the Seeger story as part of its heritage. Without a representation of the Seeger history in the unfolding story of the Whirlpool Corporation's relatively brief chronicle, the public record that Whirlpool communicates becomes nothing more than a distorted, inaccurate, and incomplete account, in essence a historical fabrication without parentage. Perhaps it was the unrecorded design of one or several corporate officials in a vain attempt to apply a stamp-of-approval to the legacy of only one participant and element of the 1955 merger of three distinctive companies.

This extensive examination of the history of the Seeger Company and the Whirlpool-Seeger and Whirlpool Corporation in St. Paul from 1902 to 1984 gives rise to a number of questions for consideration. These queries fall under the umbrella of "what if" one course of action or another had not been undertaken. My research and interviews indicate that several Seeger family members were opposed to the 1955 merger of the company with Whirlpool and RCA. These persons favored a continued independent path for Seeger, to manufacture as it had for more than fifty years its line of quality refrigeration products. The reluctant family members may have been inexperienced with regard to the management of a major corporation and the uncertainties that can result from corporate mergers. They may also have been unfamiliar with consolidations that were underway in the appliance industry in the post-World War II years. Yet they all had the perspective of nearly fifty years of close observation of the accomplishments of a company built on the leadership of their family members. They may also have been skeptical of the wisdom and benefit of deviating from such a productive past and entering an uncertain future joint venture. The course

Voices *from the* Plant Floor *at* Seeger

"Seeger's Sold the Farm"

"I thought we had bought Whirlpool according to our people in the Traffic Department. Low and behold a couple of months later, we found that Seeger's sold the farm! And I seem to remember Corporate Traffic was like a wet blanket!"

Louis F. Winkler

Demolition of the last of the former Seeger buildings on Arcade Street took place in April 1995. Craig Borck photograph. Photo courtesy of the St. Paul Pioneer Press.

of the past decades at the Seeger Company allows for this kind of speculation, but it remains impossible to determine whether Seeger would have been a stronger company after 1955 if it had remained independent. Nor is it certain that the company would have survived in the post-World War II marketplace.

Were the St. Paul employees of Whirlpool in 1984 for one reason or another intentionally mislead by the corporate decision to close the former Seeger factory? If so, why? Also, in the light of new evidence, why were such public officials as the governor of Minnesota, Rudy Perpich, and the mayor of St. Paul, George Latimer, willfully mislead regarding the closing of the East Side plant? These disturbing ethical questions remain debatable and await further examination.

What remains clear after undertaking the research and writing of the Seeger history is simply the contrasting reputations of Seeger and Whirlpool in St. Paul. The role and experience of Seeger over more than half-a-century is favorable and continues to sparkle in the community of its origin while Whirlpool's legacy is tarnished, negative, and cloaked in questionable integrity. A legacy created by its own hand.

APPENDIX A

A Partial Seeger Family Genealogy of Members Directly Associated with the Seeger Company

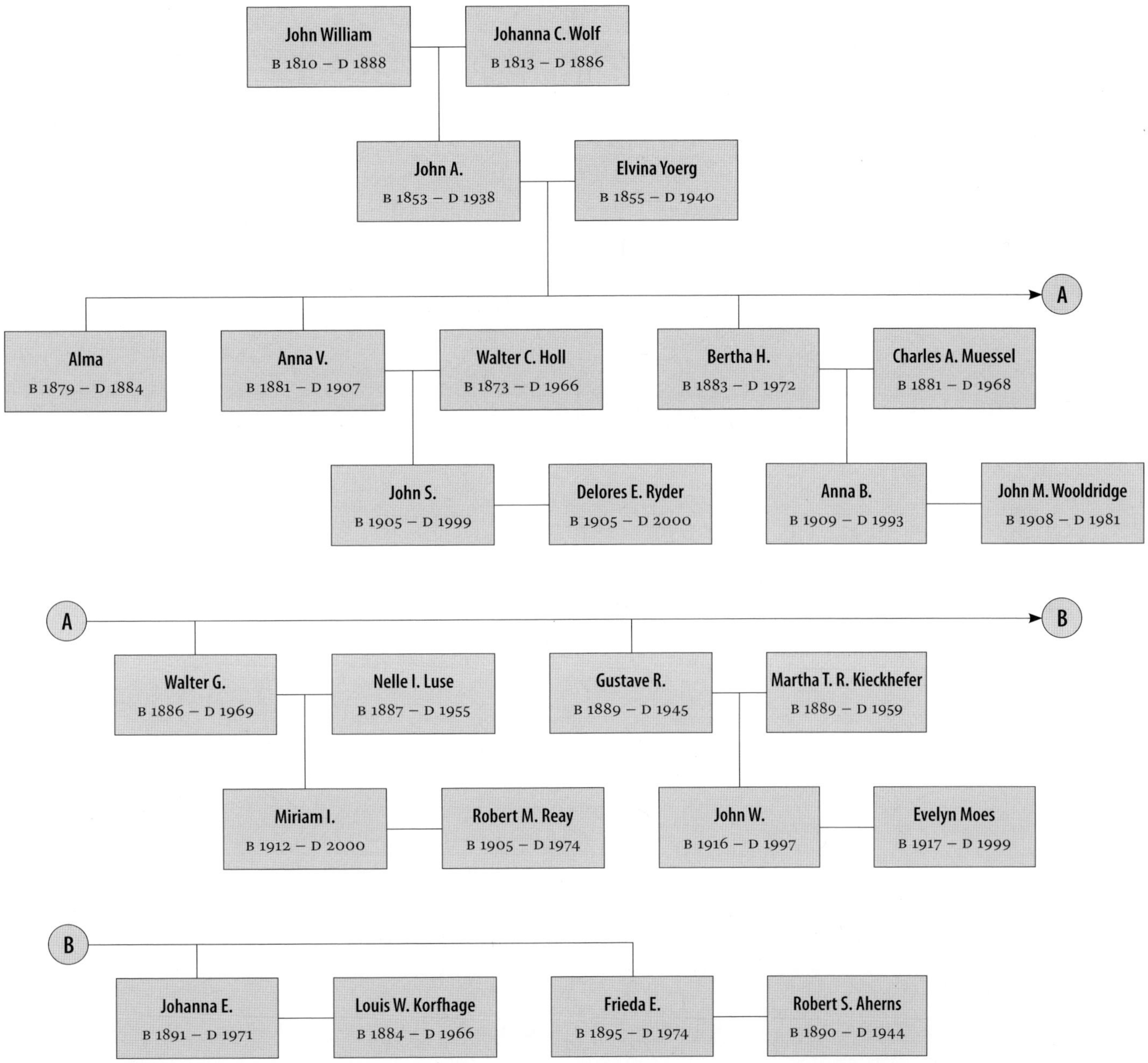

APPENDIX B

Seeger Refrigerator Company
Gross Sales and Net Income 1942–1955
(and valued in 2005 dollars)

	1942[1]	2005 DOLLARS
Gross Sales	$11,856,147	$148,770,276
Net Income	$ 945,635	$ 11,865,775
	1943	
Gross Sales	21,047,088	238,570,637
Net Income	1,138,020	12,899,539
	1944	
Gross Sales	14,156,254	151,236,959
Net Income	712,547	7,612,426
	1945[2]	
Gross Sales	14,538,089	152,720,015
Net Income	811,815	8,527,970
	1946 (Seeger-Sunbeam Corporation)	
Gross Sales	15,160,452	155,677,250
Net Income	(-87,789)	(-901,473)
	1947 (Seeger Refrigerator Company)	
Gross Sales	34,891,706	330,221,594
Net Income	1,453,005	13,751,509
	1948	
Gross Sales	62,154,289	514,195,585
Net Income	3,576,362	29,586,848

	1949	2005 DOLLARS
Gross Sales	69,155,199	530,717,386
Net Income	4,178,428	32,066,488
	1950	
Gross Sales	81,500,370	631,775,624
Net Income	5,565,535	43,142,986
	1951[3]	
Gross Sales	84,972,032	652,165,642
Net Income	4,155,175	31,891,227
	1952	
Gross Sales	88,781,380	631,513,092
Net Income	4,685,346	33,327,453
	1953	
Gross Sales	116,742,097	812,525,615
Net Income	5,169,664	35,980,888
	1954	
Gross Sales	112,564,615	777,232,459
Net Income	5,986,928	41,338,343
	1955 (seven months ended March 31, 1955 [unaudited])	
Gross Sales	44,743,694	307,407,751
Net Income	2,411,720	16,569,517

[1]Details for the Company's Gross Sales and Net Income for the years 1942–1951 are found in the 1951 *Annual Report Seeger Refrigerator Company* (St. Paul, 1951): 10 and the individual Annual Reports for subsequent years.

[2]Seeger-Sunbeam Corporation for September 1, 1944–August 31, 1945. Sales of war products represented approximately 70% of gross sales.

[3]The 1951 *Annual Report* indicates that total production was given to Sears, Roebuck and Company, Montgomery Ward & Company, the Frigidare Division of General Motors, and the Admiral Corporation.

APPENDIX C

A Geneology of Whirlpool Business Interests

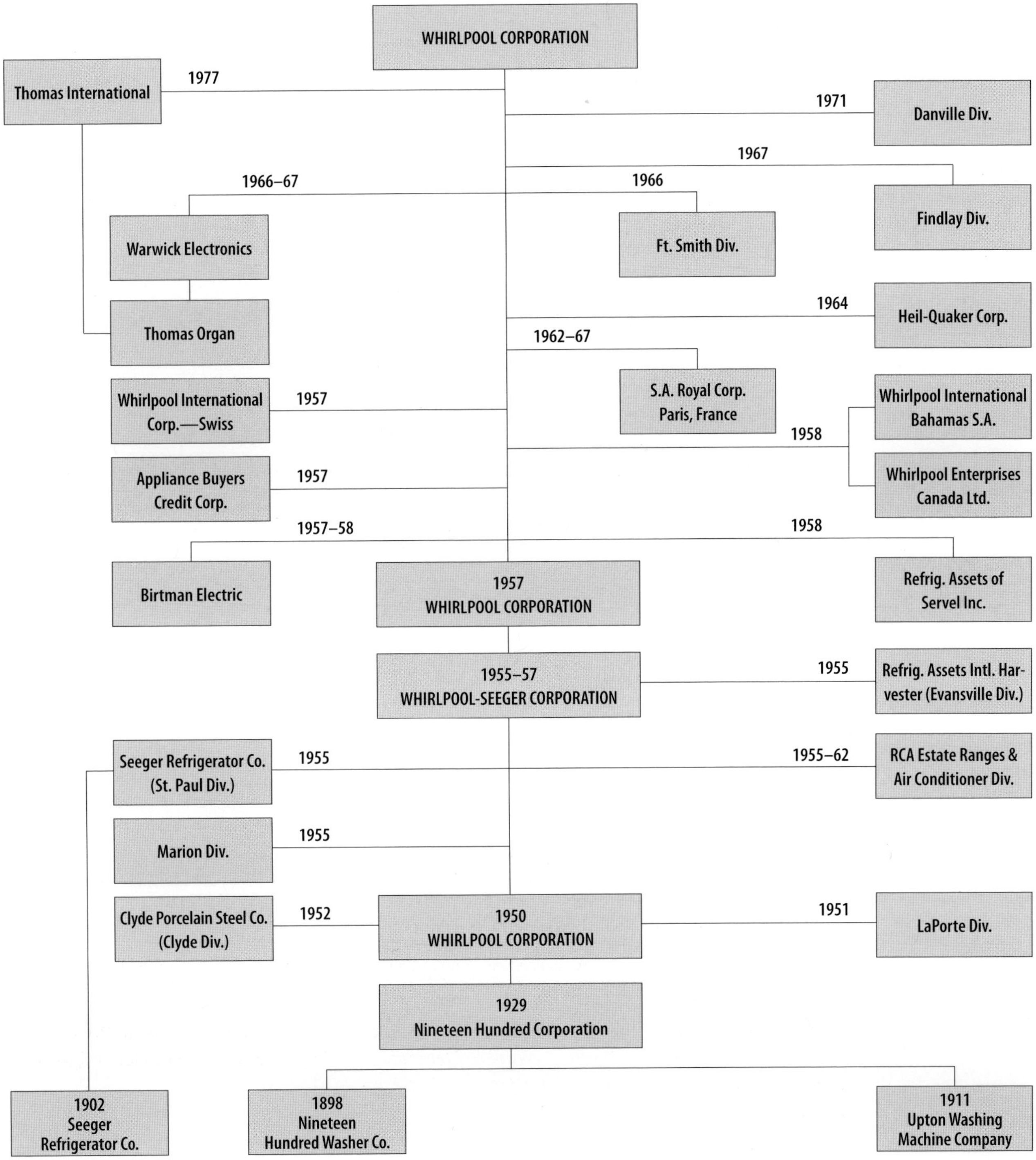

Sources

MY RESEARCH FOR THIS BOOK has been complicated by the destruction of the corporate records relating to the firm's history from 1902 to 1955. Eyewitness accounts by former employees of the Seeger Refrigerator Company and the Whirlpool Corporation informed me that the materials of that era were systematically discarded after the close of the Arcade Street factory in December 1984. The negligent act of destroying the records has not only obliterated the Seeger legacy in the broadest scope of the refrigeration industry's development, but also deprived the community of a significant element of its cultural and historical heritage.

Regretfully this book was deprived of the benefit of consulting the usual business resources essential to reconstruct and write its history such as journals, account books and ledgers, correspondence with clients and suppliers, annual reports, internal department studies and reports, and the variety of other materials generated by a large national manufacturing corporation. Nor were there printed or manuscript materials for Seeger in corporate or public archives and libraries. Consequently, to piece the chronicle together the situation required that I seek and compile information relating to its long and distinguished history from a variety of resources. The initial impetus came from Miriam Seeger Reay (Mrs. Robert M. Reay), a daughter of Walter G. Seeger and a granddaughter of John A. Seeger, who shared with me her personal collection of materials relating to her family and the company.

My search for details about the Seeger history has required a commitment of more than ten years. Over that period of time I was fortunate to meet and interview former Seeger employees who could provide me with first-hand knowledge of the company based on their having worked there. These interviews gave me an insider's look at the Seeger Company, but no single former employee is aware of all that went on at a large employer such as Seeger, even if the person was employed there for many years. In addition very few of the interviews that I was able to do were with individuals who held positions in the senior management of the company. Thus much of the material that I was able to gather focused on plant operations and very little touched on policy issues at the company.

Despite these limitations, I had the opportunity to meet with more than one hundred persons including a gentleman who began working at Seeger in 1925 and other employees whose careers with the company began in the 1930s. These former employees generously shared with me their memories and the printed materials and artifacts that they collected and preserved relating to the company's story. This book could not have been written without the assistance of Clinton Conrad, an employee of the Seeger and Whirlpool corporations from 1943 until the plant's closing in 1984. He thoughtfully and persistently amassed an extraordinary and indispensable collection of materials printed and distributed to department managers and employees that has been an essential reference source. Herbert Rhoda, in the final days of the operation of the St. Paul manufacturing facility in 1984, rescued and claimed countless important documents relating the company's manufacturing, engineering, and sales operations from the 1930s to the mid-1950s. He also rescued and preserved a portrait of the founder of the company, John A. Seeger. Many other persons generously aided my efforts and I have identified and acknowledged their kindnesses in the Preface.

The Seeger News, a magazine published three or four times a year from 1937 until the summer of 1955 was a model of corporate communication between management and workers and provided many details about the company's operations, employees, and activities. An attractive full-sized publication, it carried feature articles that described and illustrated the operations of the various departments of the firm. In addition the magazine published articles touching on employees' programs including the Men's Club, the bowling, golf, basketball, and softball teams, as well as news of the annual fishing contest. Here too, were printed personal details of employee's families—births, marriages, and deaths. During World War II and the Korean War reports were carried about employees on active duty in the military services. These wartime issues also carried lists of Seeger employees who were in uniform and singled out those who were known to have been killed in action or were being held as prisoners of war. Safety and health issues, such as the early detection of cancer and tuberculosis and the importance of contributing to local blood drives, were frequently published in the magazine.

The typical *Seeger News* was twenty-four pages long, was printed on glossy paper, and had a cover that would be printed with two inks. Each issue included a good number of photos of employees and/or events involv-

ing employees. Because the *Seeger News* seems to have been produced by employees for employees, there are very few times when it printed an article or statement written by a Seeger executive. Department managers contributed articles about the work of their departments, but overall the magazine was not a house organ that simply gave voice to company policies.

The design of the covers varied from issue to issue except for the Seeger script logo. It was always used. Based on the different numbers that have survived, the company appears to have had someone on its office staff with a journalist background who was responsible for editing each issue. As best as I can determine, no complete run of the *Seeger News* is available today. Few copies can be found in public libraries or other archival collections. The collection of the *Seeger News* that I have developed has come from family members and former employees of the company. Although this collection is missing a few issues, it remains a remarkable window on the company.

I have found it necessary to follow every lead to acquire material that would describe some aspect of the Seeger story. The search has led to me book dealers, antiques shops, and auction companies so that I might gather catalogs of the company's products, blueprints of its iceboxes and refrigerators, some of its appliance advertisements, a few of its industrial artifacts, and so forth. Some of these items are reproduced in the book. All are part of my personal Seeger collection and are not available in any public library or archive.

Within the last several years, I've used a documentary source that has been extremely valuable to me in piecing together the Seeger story. In 2002 the city of St. Paul donated its building permits from the beginning of the permitting process in 1883 to about 1970 to the Ramsey County Historical Society in St. Paul. The Society's St. Paul Building Permits Collection has provided indispensable original source information that describes the construction and estimated costs and subsequent expansions and modifications of the buildings on Arcade Street that comprised the Seeger factory. Two of the early permits are reproduced in this book so that readers who are not familiar with these documents will have some idea of what they look like and what they can tell a researcher.

The task of writing a history of a refrigerator manufacturing company requires viewing it from several perspectives. The Seeger Company was an organization that made a substantial contribution to the economy of the city of St. Paul. It also had a vital financial impact on the lives of its employees, their families and dependents, and the larger community. The Seeger Company was also a social organization with obligations and responsibilities. For both its management and its employees, their joint response to the demands of the marketplace for appliances and their ability

to cooperate in meeting these demands determined the success of the company's operations.

The chapter endnotes that follow indicate the source materials that I have uncovered and used in writing this book. They also illuminate, describe, and confirm specific elements of *From Arcade Street to Main Street: A History of the Seeger Refrigerator Company, 1902–1984.*

Notes

1. 1856: The Seeger Family Arrives in Minnesota

1. For an extensive discussion of the controversy surrounding William Seeger during his few months of service as treasurer of the state of Minnesota, see Frank R. Holmes, *Minnesota in Three Centuries* (St. Paul, 1908), 4:54–56; also, William Watts Folwell, *A History of Minnesota* (St. Paul, 1926), 3:357–62. I am grateful to Scott Seeger Bell for suggesting that these works should be read along with Theodore Dreiser's fascinating novel, *The Financier*, which chronicles the financial practices of city and state government officials after the Civil War. Because much new information is now available regarding partisan political financial interests and conflicts in state legislatures of the period, the Seeger incident in Minnesota should be revisited for a more complete assessment.

2. Ramsey County Historical Society, St. Paul Building Permits Collection, St. Paul Building Permits no. 8791, October 1, 1886 and no. 13178, October 5, 1887.

3. For details regarding James J. Hill's loan to the Bohn Manufacturing Company and its chief officers and his subsequent legal action to recover monies owed him, see the records of the St. Paul law firm of Davis, Kellogg and Severance, Minnesota Historical Society, file number 2302; MSS James J. Hill Papers, Notes and Bills Receivables, 1881–1917, numbers: 529-31, 552, 617, 756, 841–42, James J. Hill Reference Library, St. Paul, Minnesota.

4. For details regarding the Bohn Manufacturing Company's proposal to provide the interior woodwork for the state Capitol building, signed by John A. Seeger as president, see the State Capitol Board of Commissioner Records, Minnesota Historical Society, box number 3; State Capitol Board of Commissioners Minute Book, 315, 325; *St. Paul Pioneer Press*, June 4, 1902. Also the Cass Gilbert Collection, Drawings for the Minnesota State Capitol, Minnesota Historical Society, Plans 281–87, 290–91. For correspondence between Bohn Manufacturing Company and Butler Brothers Construction Company, see the Davis, Kellogg and Severance Records, Minnesota Historical Society, file number 2355. I am grateful to Carolyn Kompelien, manager of the Capitol site, for sharing with me her knowledge of the building's original interior woodwork and for providing me with an extensive tour of the Capitol to view firsthand the work of John A. Seeger's firm. For details relating to the legal proceedings surrounding the liquidation of the Bohn Manufacturing Company, see the Davis, Kellogg and Severance Records, file numbers 2302, 2355, 2380, and 2387.

5. See *The Seeger News*, "Forty Years of Progress," June 1946, vol. 10, no. 2.

6. See the Davis, Kellogg and Severance records, file numbers 3047, 3048, and 3049 for extensive details regarding the lawsuit. *The G.F. Quinn Refrigerator Co., Catalogue 1896* (Portland, 1896): 3. *The Federal Reporter,* June–August 1910 (St. Paul, 1910): 567–77.

John H. White, *The Great Yellow Fleet: A History of American Railroad Refrigerator Cars* (San Marino, Calif., 1986): 59–60.

7. The Seeger Refrigerator Company's acquisition of the former Bohn Manufacturing Company property on Arcade Street is documented in the Ramsey County Register of Deeds Records, May 11, 1910, record numbers 405178, 405179. See also *Seeger News,* 10 (June 1946): 4.

2. From Arcade Street to Main Street: Iceboxes and Refrigerators

1. *Seeger Refrigerator Company Catalog* (St. Paul, 1920): 30.

2. Seeger Refrigerator Company, *The Original Siphon Refrigerator,* Catalog 18, Supplement (St. Paul, 1914).

3. *General Electric Refrigerators Salesman's Handbook,* 3rd edition (Cleveland, 1928): 50–58.

4. Miriam Seeger Reay, Family Papers.

5. Seeger Refrigerator Company, Minute Book of the Board of Directors, October 15, 1926.

6. Ibid., October 20, 1927.

7. Ibid., October 18, 1928; October 28, 1929.

8. *Catalog No. 45A, Seeger Original Siphon Refrigerators for Every Purpose, Seeger Refrigerator Company* (St. Paul, 1926). *Cabinets by Seeger for Electrical Refrigeration* (St. Paul, 1926).

9. *Catalog No. 50, Cabinets by Seeger, St. Paul, For the American Home, Seeger Refrigerator Company* (St. Paul, 1930).

10. *Catalog No. 49, Commercial Cabinets by Seeger, St. Paul, Seeger Refrigerator Company*-(St. Paul, 1930).

11. Ramsey County Historical Society Mss., St. Paul Building Permits, Nos. 6108, Nov. 5, 1923; 10094, July 15, 1924; 15640, June 5; 16025, June 25; 16152, June 30; 16215, July 3; 16331, Aug. 5; 17301, Aug. 27, 1925; 22198, June 28; 22503, July 15; 23029, Aug. 10, 1926; 29094, Oct. 10, 1927. According to the Building Permits, the anticipated cost of the improvements totaled $205,500, or $2,225,646 in 2005 inflation-adjusted dollars.

12. A physical description of the various available household and commercial cabinets that Seeger manufactured with Fairbanks Morse electrical refrigeration units installed in them is found in *Kelvinator Sales Data, Kelvinator Inc.* (Detroit, 1927). The Collection of James B. Bell.

3. Refrigeration Progress: An Era of Residential and Commercial Change

1. Charles Singer, E. J. Holmyard, A. R. Hall, and Trevor I. Williams, eds., *A History of Technology* (Oxford, 1958), 5:45–51; Trevor I. Williams, *A History of Technology: The Twentieth Century c. 1900–c. 1950* (Oxford, 1978), 2:1144–46; and Oscar Edward Anderson, *Refrigeration in America: A History of New Technology and its Impact* (Princeton, 1953).

2. A 1912 Seeger Catalog, describes the construction, insulation, and operation of the seventeen products then manufactured by the company. *Catalog No. 10, Seeger Siphon Refrigerators, Manufactured by the Seeger Refrigerator Company* (St. Paul, 1912).

3. W. R. Woolrich, *The Men Who Created Cold* (New York, 1967), 20–29, 58–124. Anderson, *Refrigeration in America*: 71–96.

4. Refrigeration is the process of removing heat from an enclosed space, or from a substance, to lower its temperature. One type of refrigerator uses the evaporation of a liquid to absorb heat. The liquid, or refrigerant, used in a refrigerator evaporates at an

extremely low temperature, creating freezing temperatures inside the refrigerator. This is all based on the following physics: a liquid is rapidly vaporized; the quickly expanding vapor requires kinetic energy and draws the energy needed from the immediate area, which thus loses energy and becomes cooler. Cooling caused by the rapid expansion of gases is the primary means of refrigeration today.

Mechanical refrigeration works on the principle that liquids absorb heat when they vaporize. You can demonstrate this by wetting your hands and waving them rapidly. The water evaporates quickly and causes a cooling sensation by lowering the skin temperature. A fan cools you because it evaporates the natural moisture on your skin.

Mechanical refrigeration includes three principal systems. They are (1) compression, (2) absorption, and (3) steam-jet. The electric refrigerator is a compression-type refrigeration unit powered by an electric motor. A home electric refrigerator consists of five basic parts: (1) the receiver, (2) the refrigerant-control device, (3) the evaporator, (4) the compressor, and (5) the condenser.

At the beginning of the refrigeration cycle, the refrigerant leaves the *receiver* (storage tank) under high pressure. The refrigerant travels through pipes to the *refrigerant-control device*. This mechanism reduces the pressure of the refrigerant as it enters the *evaporator*. The evaporator is the coldest spot in the refrigerator and serves as the freezing unit. It consists of pipes or coils on the walls or sides of the cabinet, or surrounding the ice-tray compartment. At a low pressure, the liquid refrigerant evaporates inside these coils and absorbs heat. This causes refrigeration to take place. The compressor pumps the refrigerant from the freezing unit as a vapor, and raises its pressure. It then discharges high-pressure gas into the air-cooled condenser. There the gas loses the heat it gained in the evaporator and condenses into a liquid at the high pressure, which flows back to the storage tank.

Compression and absorption systems refrigerate by changing a refrigerant from a liquid to a gas and back to a liquid again. These repeated operations make up the refrigeration cycle. In a compression system, a compressor brings about the refrigeration cycle. This system is widely used in industry and in most home electric refrigerators. In the absorption system, the refrigeration cycle is caused by the direct application of heat from gas, steam, or some other source. All home and camper gas refrigerators, and some industrial units, use the absorption system. Jacob Perkins, a Massachusetts inventor, developed the first compression machine in 1834. During the 1850s, Ferdinand Carré, a French engineer, developed the first absorption system using ammonia. Carl von Linde of Germany introduced the first successful compression system using ammonia between 1873 and 1875.

5. Anderson, *Refrigeration in America*, 14–35.

6. For an interesting account of the nineteenth century development of this industry in Boston, Massachusetts, see Gavin Weightman, *The Frozen Water Trade: How Ice From New England Lakes Kept the World Cool* (London, 2001).

4. 1928: An Alliance with Sears Begins

1. The Seeger Refrigerator Company was a preferred and common stockholder of the Servel Corporation of Evansville, Indiana (Seeger Refrigerator Company, Minute Book of the Board of Directors, July 24, 1928). Because Walter G. Seeger had a more than thirty-year association with Servel executives, it would be interesting to know the details of his involvement with Servel during his chairmanship of the board of directors of the Whirlpool-Seeger in 1957.

2. Tennessee Furniture Corporation, *Coldspot Refrigerator Cabinets: General Information and Service Instructions* (Chattanooga, ca. 1926).

3. Sunbeam Electric Manufacturing Company Catalog (Evansville, 1928). See a discussion of the structure and operation of the 1938 Coldspot Refrigerator unit in Master Service Manuals, Household Refrigeration (Detroit, 1938): 443–55.

4. See *The Seeger News,* 12, no. 2 (Winter 1949); Ibid., 15, no. 3 (Fall 1952).

5. See *The Seeger News,* 16, no. 3 (Fall 1953).

6. See *The Seeger News,* 15, no. 4 (Winter 1952).

5. Parts and Tools for Production

1. See *The Seeger News,* 11, no. 2 (Spring 1948).

2. I am indebted to articles in *The Seeger News* for information about the complex operations of this department, including vol. 6, no. 2 (May 1942); Ibid., 12, no. 2 (Summer 1949); Ibid., 16, no. 2 (Summer 1953) and 16, no. 4 (Winter 1953).

3. See *The Seeger News,* 16, no. 1 (Spring 1953).

4. See *The Seeger News,* 15, no. 2 (Summer 1952).

6. Porcelain Enamel Finishing and Stocked Parts

1. See *The Seeger News,* 10, no. 1 (March 1946); Ibid., 14, no. 3 (Fall 1951).

2. See *The Seeger News,* 10, no. 2 (June 1946).

3. See *The Seeger News,* 13, no. 2 (Fall 1950).

7. 1934–1941: The Impact of the Sears *Coldspot Super Six*

1. The only documentation that has survived of the specific changes that the Seeger management made in the Arcade Street plant between 1934 and 1941 can be found in the St. Paul Building Permits No. 38,610 (September 25, 1934); No. 40,120 (October 31, 1934; and No. 49,842 (August 26, 1935); and No. 86,640 (July 21, 1941), St. Paul Building Permits Collection, Ramsey County Historical Society, St. Paul, Minnesota.

2. John A. Seeger, St. Paul, Minnesota, to the Dent Hardware Company, February 16, 17, 1911. *Catalogue of Refrigerator Hardware, Volume C, Manufactured by The Dent Hardware Company* (Fullerton, 1910).

3. For descriptions and illustrations of the Seeger manufactured cabinets for Sears, Roebuck & Co. and Montgomery Ward between 1930–46 and 1935–46 respectively, see Edythe Polster and J. Louis Albert, eds., *Standard Refrigerator Trade-In & Dealer Guide, 1946* (New York, 1946): 9–18, 137–45.

4. *New York Times,* April 15, 1994, A-29; Raymond Loewy, *Never Leave Well Enough Alone* (New York, 1951): 125–130. *The Designs of Raymond Loewy* (Washington, D.C., 1975): 37. Angela Schonberger ed., *Raymond Loewy: Pioneer of American Industrial Design* (Munich, 1990): 80–81, 146, 232.

5. See *The Seeger News,* 6, no. 3 (August 1942); Ibid., 7, no. 1 (June 1943); Ibid., 9, no. 3 (October, 1945); Ibid., 9, no. 4 (December 1945); Ibid., 12, no. 3 (Summer 1948); Ibid., 15, no. 2 (Summer 1951).

8. 1942–1945: A Partner in National Defense with Wartime Production

1. See *The Seeger News,* 6, no. 2 (May, 1942) [wartime storage refrigerators]; Ibid., [ammunition lockers]; Ibid., 6, no. 3 (August, 1942) [ammunition boxes]; Ibid., 7, no. 1 (June 1943) [flexible ammunition chutes]; Ibid. [96-foot coolers]; Ibid. [pressure cookers]; Ibid. [Masonite refrigerators]; Ibid., 7, no. 2 (December 1943) [parachute flairs]; Ibid. [bomb racks]; Ibid., 8, no. 2 (September 1944) [B-29 Superfortress bomb racks];

Ibid., 8, no. 3 (December 1944) [E-Award]; Ibid., 15, no. 2 (Summer 1952) [jet fuel tanks].

9. 1945–1955: A Decade of Robust Productivity

1. For details regarding the merger of Seeger-Sunbeam Corporation, I am indebted to the Thomas Reed Powell Papers, 1905–1955, Harvard Law School Library, Cambridge, Mass., Folder E5, which includes a copy of the Seeger-Sunbeam Purchase Agreement, June 12, 1945. Additional information on the Seeger acquisition of the Sunbeam Corporation is in *Investor's Reader,* 8, no. 8 (April 16, 1947): 14.

2. *Seeger-Sunbeam Corporation Annual Report, 1946* (St. Paul, 1946).

3. Remarks of Walter G. Seeger on the history of the Seeger Refrigerator Company to the Board of Directors, Northwest Bancorporation, April 23, 1948.

4. The expansion program is documented in St. Paul Building Permits No. 122,580 (August 16, 1945); No. 124,708 (September [no day identified], 1945); No. 127,112 (December 3, 1945); and No. 129,085 (February 8, 1946), St. Paul Building Permits Collection, Ramsey County Historical Society, St. Paul, Minnesota.

5. See *The Seeger News,* 16, no. 1 (Spring 1953); Ibid., 17, no. 1 (Spring 1954); Ibid., 17, no. 2 (Summer 1954).

6. See *Seeger News,* 18, no. 1 (Spring 1955); Ibid., 18, no. 2 (Summer 1955).

7. *Northern States Power Magazine,* (March 1955): 3–5.

8. See *Seeger News,* 11, no. 4 (Fall 1948).

9. See Appendix B for the Seeger Company's annual gross sales and net income for the years 1942–1955.

10. The 1955 Merger with the Whirlpool Corporation and the RCA Appliance Division

1. Minutes of Special Meeting of Board of Directors of Seeger Refrigerator Company, May 3, 1955.

2. Walter G. Seeger, chairman, John S. Holl, president, Seeger Refrigerator Company to Shareholders, St. Paul, Minnesota, July 16, 1955.

3. Ibid.

4. Ibid.

5. Ibid.

6. Letter of Walter G. Seeger, chairman, and John S. Holl, president, Seeger Refrigerator Company to Shareholders, St. Paul, Minnesota, August 15, 1955, 7.

7. Ibid., 5.

8. Ibid., 6.

9. Ibid., 8.

10. Ibid.

11. Ibid.

12. Ibid.

13. Ibid., 4.

14. Ibid., 5.

15. Ibid., 5. *1955 RCA Annual Report* (New York, 1956): 4.

16. Letter of Walter G. Seeger, chairman, and John S. Holl, president, Seeger Refrigerator Company to Shareholders, St. Paul, Minnesota, August 15, 1955, 8-9. For a discussion of the terms of the Preferred Stock, see the letter "To the Stockholders of Whirlpool-Seeger Corporation, January, 6, 1956," Whirlpool-Seeger Corporation, St. Joseph, Michigan.

11. 1955–1984: The Merged Corporation

1. *Whirlpool-Seeger* (St. Joseph, 1955); "A Progress Report: One Year of Whirlpool-Seeger," *Appliance Manufacturer*, 5, no. 1 (January 1957): 40–105; *Investor's Reader*, 28, no. 5 (March 6, 1957): 13–15; and *Forbes Business & Finance*, (February 1, 1963): 18–22.

2. *Whirlpool Corporation, Annual Report, 1970.*

3. This highly productive association collapsed under curious circumstances during the years when Whirlpool was led by John Platts and Jack Sparks. The seven years (1972-76) of litigation that marked the termination of the contractual relationship culminated in a trial at which a federal jury in New York City found that Whirlpool and Sears, Roebuck and Company were guilty of a conspiracy to restrain trade in violation of the Sherman Antitrust Act. This case (*Oreck Corp. v. Whirlpool Corp.*, 639 F.2d 75 (2d Cir. 1980)) has become a landmark legal proceeding and the decision is widely studied in law schools.

4. *Inside Whirlpool News*, August 27, 1968. Collection of Clinton Conrad.

5. The two presidents and chairmen of the Whirlpool Corporation during this period, John Platts and Jack Sparks, both noted in their *Who's Who in America* entries that they had attended the University of Toledo. After an inquiry to the University of Toledo's president, Dr. Dan Johnson, Ms. Julie Quinoñez of the University's registrar's office has informed me that Platts attended the university during the academic year 1938–39 for one semester. The University of Toledo has no record that Sparks was ever enrolled at the institution. There are other details in the *Who's Who* entries of Platts and Sparks that are at variance with the facts.

6. *Inside Whirlpool News*, September 4, 1968. Collection of Clinton Conrad.

7. Ibid.

8. Wright Anderson Williams, "Deindustrialization in East Saint Paul: The Whirlpool Plant Closing of 1984," undergraduate research paper at Macalester College, 1996.

9. The saga continues with massive layoffs at Whirlpool freezer and refrigerator manufacturing centers. In early October 2006, Whirlpool announced that 500 workers at the Evansville freezer factory would be laid off due to expected lower demand for the appliance but 400 persons would be added to the work force at the Amana, Iowa, freezer plant. At the Fort Smith, Arkansas, refrigerator factory 1,640 employees were terminated and production of the appliance shifted to and increased at the Ramos Arizpe, Mexico, plant.

10. "Whirlpool to Lay off 300 Employees," *St. Paul Dispatch*, August 25, 1982.

11. No copy of this "study" has been found in the corporate records archives of the Whirlpool Corporation in Benton Harbor, Michigan. See also note 16.

12. Don Clark, "Whirlpool Studies Plant Need," May 7, 1983, sec. C1.

13. *St. Paul Pioneer Press*, July 16, 1983, sec. A1.

14. Don Clark and Brian Bonner, "Whirlpool to Consider Move South," *St. Paul Pioneer Press*, August 7, 1983, sec. A5.

15. Brian Bonner, "Job Losses Leave Employees with Grim Outlook," *St. Paul Pioneer Press*, July 16, 1983, sec. A5.

16. Ibid., April 20, 1984, sec. B16. No copies of either "study" has been found in the corporate archives of the Whirlpool Corporation in Benton Harbor, Michigan, nor do the Minutes of the board of directors of the corporation carry any notice of a "study" of either the prospective or impending shutdown of the St. Paul facilities. The last division vice president of the St. Paul operations, Donald Casey, has indicated that he did provide cost estimates for relocating tooling and equipment to both Danville and Evansville but that he did not see a full report of the study. Because a decision by corporate officials to close a major manufacturing facility is a significant policy issue, presumably

requiring due deliberation and approval of the board of directors, it is puzzling that copies of the studies to close the Cottage Grove and Arcade Street factories have not been found. As a historian, or any responsible investigator would suggest, a question arises: Was a written study ever developed to determine the fate of either manufacturing facility? Or, alternatively, were such references to the two studies merely a public relations gesture on the part of the Whirlpool Corporation to mask the decisions of corporate officers? These questions remain open for consideration today.

17. Ibid., April 27, 1984, sec. A1, 4.

18. Ibid., December 23, 1984, sec. A1; December 28, 1984, sec. C3; December 29, 1984, sec. A1.

19. Alfred King, "Whirlpool Leaving St. Paul," *St. Paul Pioneer Press*, April 27, 1984, sec. A4.

20. "Don Casey Letter." *Twelve O'Clock High*, special issue, April 26, 1984, Collection of Clinton Conrad.

21. Karl Karlson, "Workers Take News of Shutdown Hard," *St. Paul Pioneer Press*, April 27, 1984, sec. 4A.

22. Jacqui Banaszynskiv, "Whirlpool Severance Offered," *St. Paul Pioneer Press*, July 28, 1984, sec. A3.

23. Ibid., December 23, 1984, sec. A1. The number of years that the 345 employees with twenty of more years of service was as follows: 20 years—8; 21 through 24 years—15; 25 years—15; 26 through 29 years—72; 30 through 34 years—137; 35 through 39 years—95; and 40 years and over—3. "Don Casey Letter," *Twelve O'Clock High*, October 26, 1984. Collection of Clinton Conrad.

24. *St. Paul Pioneer Press*, December 23, 1984, sec. A1.

25. Ibid., December 28, 1984, sec. C1.

26. Ibid., December 29, 1984, sec. A1.

27. Ibid., July 23, 1989, sec. B1.

28. Ibid., November 26, 1993, sec. E1; *Wall Street Journal*, January 27, 1994, sec. B4; *Wall Street Journal*, July 13, 1994, sec. A10; and *Barron's*, September 26, 1994, 17–18.

Index

N

O

P

R

S

T